Principles and Methods in Plant Molecular Biology

Principles and Methods in Plant Molecular Biology

Dr. R.A.S. Tomar

Editor

KOROS PRESS LIMITED

London, UK

Principles and Methods in Plant Molecular Biology

© 2012

Printed in 2017 for Sale in the Indian Subcontinent

Published by
Koros Press Limited
3 The Pines, Rubery B45 9FF, Rednal,
Birmingham, United Kingdom

Tel.: +44-7826-930152
Email: info@korospress.com
www.korospress.com

ISBN: 978-1-78163-025-9

Editor: Dr. R.A.S. Tomar

Printed in UK

British Library Cataloguing in Publication Data
A CIP record for this book is available from the British Library

10 9 8 7 6 5 4 3 2 1

Contents

Antiparallel Strands of DNA • DNA Damage, Repair, and Recombination • DNA Damage • Global Response to DNA Damage • DNA Repair and Aging • DNA Repair and Evolution • Senescence • Transcription • Major Steps • Transcription Factories • Ribonucleic Acid • Comparison with DNA • Key Discoveries in RNA Biology • Transcription and Processing of RNA • Gene Transcription: DNA '! RNA • Types of RNA • RNA Processing: Pre-mRNA '! mRNA • Messenger RNA • Ribozyme • Nuclear Polyadenylation • Cytoplasmic Polyadenylation • Messenger RNA Decapping • Open Reading Frame • The RNA-induced Silencing Complex

Preface

Since the late 1950s and early 1960s, molecular biologists have learned to characterize, isolate, and manipulate the molecular components of cells and organisms. These components include DNA, the repository of genetic information; RNA, a close relative of DNA whose functions range from serving as a temporary working copy of DNA to actual structural and enzymatic functions as well as a functional and structural part of the translational apparatus; and proteins, the major structural and enzymatic type of molecule in cells. One of the most basic techniques of molecular biology to study protein function is expression cloning. In this technique, DNA coding for a protein of interest is cloned (using PCR and/or restriction enzymes) into a plasmid (known as an expression vector). This plasmid may have special promoter elements to drive production of the protein of interest, and may also have antibiotic resistance markers to help follow the plasmid.

This plasmid can be inserted into either bacterial or animal cells. Introducing DNA into bacterial cells can be done by transformation (via uptake of naked DNA), conjugation (via cell-cell contact) or by transduction (via viral vector). Introducing DNA into eukaryotic cells, such as animal cells, by physical or chemical means is called transfection. Several different transfection techniques are available, such as calcium phosphate transfection, electroporation, microinjection and liposome transfection. DNA can also be introduced into eukaryotic cells using viruses or bacteria as carriers, the latter is sometimes called bactofection and in particular uses Agrobacterium tumefaciens. The plasmid may be integrated into the genome, resulting in a stable transfection, or may remain independent of the genome, called transient transfection.

In either case, DNA coding for a protein of interest is now inside a cell, and the protein can now be expressed. A variety of systems, such as inducible promoters and specific cell-signaling factors, are available to help express the protein of interest at high levels. Large quantities of a protein can then be extracted from the bacterial or eukaryotic cell. The protein can be tested for enzymatic activity under a variety of situations, the protein may be crystallized so its tertiary structure can be studied, or, in the pharmaceutical industry, the activity of new drugs against the protein can be studied.

The book is a meticulously organized and useful both for teaching and for reference. It is intended to serve plant biology and related disciplines, ranging from molecular biology and biotechnology to biochemistry.

—Editor

1

Fundamentals of Molecular Biology

Molecular biology is the branch of biology that deals with the molecular basis of biological activity. This field overlaps with other areas of biology and chemistry, particularly genetics and biochemistry. Molecular biology chiefly concerns itself with understanding and the interactions between the various systems of a cell, including the interactions between the different types of DNA, RNA and protein biosynthesis as well as learning how these interactions are regulated.

Writing in *Nature* in 1961, William Astbury described molecular biology as not so much a technique as an approach, an approach from the viewpoint of the so-called basic sciences with the leading idea of searching below the large-scale manifestations of classical biology for the corresponding molecular plan. It is concerned particularly with the *forms* of biological molecules and [...] is predominantly three-dimensional and structural—which does not mean, however, that it is merely a refinement of morphology. It must at the same time inquire into genesis and function.

Relationship to other Biological Sciences

Researchers in molecular biology use specific techniques native to molecular biology, but increasingly combine these with techniques and ideas from genetics and biochemistry. There is not a defined line between these disciplines. The figure above is a schematic that depicts one possible view of the relationship between the fields:

- *Biochemistry* is the study of the chemical substances and vital processes occurring in living organisms. Biochemists focus heavily on the role, function, and structure of biomolecules.

The study of the chemistry behind biological processes and the synthesis of biologically active molecules are examples of biochemistry.

- *Genetics* is the study of the effect of genetic differences on organisms. Often this can be inferred by the absence of a normal component (e.g. one gene). The study of "mutants" – organisms which lack one or more functional components with respect to the so-called "wild type" or normal phenotype. Genetic interactions (epistasis) can often confound simple interpretations of such "knock-out" studies.

- *Molecular biology* is the study of molecular underpinnings of the processes of replication, transcription, translation, and cell function. The central dogma of molecular biology where genetic material is transcribed into RNA and then translated into protein, despite being an oversimplified picture of molecular biology, still provides a good starting point for understanding the field. This picture, however, is undergoing revision in light of emerging novel roles for RNA.

Much of the work in molecular biology is quantitative, and recently much work has been done at the interface of molecular biology and computer science in bioinformatics and computational biology. As of the early 2000s, the study of gene structure and function, molecular genetics, has been among the most prominent sub-field of molecular biology.

Increasingly many other loops of biology focus on molecules, either directly studying their interactions in their own right such as in cell biology and developmental biology, or indirectly, where the techniques of molecular biology are used to infer historical attributes of populations or species, as in fields in evolutionary biology such as population genetics and phylogenetics. There is also a long tradition of studying biomolecules "from the ground up" in biophysics.

Techniques of Molecular Biology

Since the late 1950s and early 1960s, molecular biologists have learned to characterize, isolate, and manipulate the molecular components of cells and organisms. These components include DNA, the repository of genetic information; RNA, a close relative of DNA whose functions range from serving as a temporary working copy of DNA to actual structural and enzymatic functions as well as a functional and structural part of the translational apparatus; and proteins, the major structural and enzymatic type of molecule in cells.

Expression Cloning

One of the most basic techniques of molecular biology to study protein function is expression cloning. In this technique, DNA coding for a protein of interest is cloned (using PCR and/or restriction enzymes) into a plasmid (known as an expression vector). This plasmid may have special promoter elements to drive production of the protein of interest, and may also have antibiotic resistance markers to help follow the plasmid.

This plasmid can be inserted into either bacterial or animal cells. Introducing DNA into bacterial cells can be done by transformation (via uptake of naked DNA), conjugation (via cell-cell contact) or by transduction (via viral vector). Introducing DNA into eukaryotic cells, such as animal cells, by physical or chemical means is called transfection. Several different transfection techniques are available, such as calcium phosphate transfection, electroporation, microinjection and liposome transfection. DNA can also be introduced into eukaryotic cells using viruses or bacteria as carriers, the latter is sometimes called bactofection and in particular uses Agrobacterium tumefaciens. The plasmid may be integrated into the genome, resulting in a stable transfection, or may remain independent of the genome, called transient transfection.

In either case, DNA coding for a protein of interest is now inside a cell, and the protein can now be expressed. A variety of systems, such as inducible promoters and specific cell-signaling factors, are available to help express the protein of interest at high levels. Large quantities of a protein can then be extracted from the bacterial or eukaryotic cell. The protein can be tested for enzymatic activity under a variety of situations, the protein may be crystallized so its tertiary structure can be studied, or, in the pharmaceutical industry, the activity of new drugs against the protein can be studied.

Polymerase Chain Reaction (PCR)

The polymerase chain reaction is an extremely versatile technique for copying DNA. In brief, PCR allows a single DNA sequence to be copied (millions of times), or altered in predetermined ways. For example, PCR can be used to introduce restriction enzyme sites, or to mutate (change) particular bases of DNA, the latter is a method referred to as "Quick change". PCR can also be used to determine whether a particular DNA fragment is found in a cDNA library. PCR has many variations, like reverse transcription PCR (RT-PCR) for amplification of RNA, and, more recently, real-time PCR (QPCR) which allow for quantitative measurement of DNA or RNA molecules.

Gel Electrophoresis

Gel electrophoresis is one of the principal tools of molecular biology. The basic principle is that DNA, RNA, and proteins can all be separated by means of an electric field. In agarose gel electrophoresis, DNA and RNA can be separated on the basis of size by running the DNA through an agarose gel. Proteins can be separated on the basis of size by using an SDS-PAGE gel, or on the basis of size and their electric charge by using what is known as a 2D gel electrophoresis.

Macromolecule Blotting and Probing

The terms *northern, western* and *eastern* blotting are derived from what initially was a molecular biology joke that played on the term *Southern blotting*, after the technique described by Edwin Southern for the hybridisation of blotted DNA. Patricia Thomas, developer of the RNA blot which then became known as the *northern blot* actually didn't use the term. Further combinations of these techniques produced such terms as *southwesterns* (protein-DNA hybridizations), *northwesterns* (to detect protein-RNA interactions) and *farwesterns* (protein-protein interactions), all of which are presently found in the literature.

Southern Blotting

Named after its inventor, biologist Edwin Southern, the Southern blot is a method for probing for the presence of a specific DNA sequence within a DNA sample. DNA samples before or after restriction enzyme digestion are separated by gel electrophoresis and then transferred to a membrane by blotting via capillary action. The membrane is then exposed to a labeled DNA probe that has a complement base sequence to the sequence on the DNA of interest. Most original protocols used radioactive labels, however non-radioactive alternatives are now available. Southern blotting is less commonly used in laboratory science due to the capacity of other techniques, such as PCR, to detect specific DNA sequences from DNA samples. These blots are still used for some applications, however, such as measuring transgene copy number in transgenic mice, or in the engineering of gene knockout embryonic stem cell lines.

Northern Blotting

The northern blot is used to study the expression patterns of a specific type of RNA molecule as relative comparison among a set of different samples of RNA. It is essentially a combination of denaturing RNA gel electrophoresis, and a blot. In this process RNA is separated based on size and is then transferred to a membrane that is then probed with a labeled complement of a sequence of interest.

The results may be visualized through a variety of ways depending on the label used; however, most result in the revelation of bands representing the sizes of the RNA detected in sample. The intensity of these bands is related to the amount of the target RNA in the samples analysed. The procedure is commonly used to study when and how much gene expression is occurring by measuring how much of that RNA is present in different samples. It is one of the most basic tools for determining at what time, and under what conditions, certain genes are expressed in living tissues.

Western Blotting

Antibodies to most proteins can be created by injecting small amounts of the protein into an animal such as a mouse, rabbit, sheep, or donkey (polyclonal antibodies) or produced in cell culture (monoclonal antibodies). These antibodies can be used for a variety of analytical and preparative techniques.

In western blotting, proteins are first separated by size, in a thin gel sandwiched between two glass plates in a technique known as SDS-PAGE (sodium dodecyl sulfate polyacrylamide gel electrophoresis). The proteins in the gel are then transferred to a PVDF, nitrocellulose, nylon or other support membrane. This membrane can then be probed with solutions of antibodies. Antibodies that specifically bind to the protein of interest can then be visualized by a variety of techniques, including colored products, chemiluminescence, or autoradiography. Often, the antibodies are labeled with enzymes. When a chemiluminescent substrate is exposed to the enzyme it allows detection. Using western blotting techniques allows not only detection but also quantitative analysis.

Analogous methods to western blotting can be used to directly stain specific proteins in live cells or tissue sections. However, these *immunostaining* methods, such as FISH, are used more often in cell biology research.

Eastern Blotting

Eastern blotting technique is to detect post-translational modification of proteins. Proteins blotted on to the PVDF or nitrocellulose membrane are probed for modifications using specific substrates.

Arrays

A DNA array is a collection of spots attached to a solid support such as a microscope slide where each spot contains one or more single-stranded DNA oligonucleotide fragment. Arrays make it possible to put down a large quantity of very small (100 micrometre diameter) spots on a single slide. Each spot has a DNA fragment molecule that is complementary to a single DNA sequence (similar to Southern blotting). A variation of this technique allows the gene expression of an organism at a particular stage in development to be qualified (expression profiling). In this technique the RNA in a tissue is isolated and converted to labeled cDNA.

This cDNA is then hybridized to the fragments on the array and visualization of the hybridization can be done. Since multiple arrays can be made with exactly the same position of fragments they are particularly useful for comparing the gene expression of two different tissues, such as a healthy and cancerous tissue. Also, one can measure what genes are expressed and how that expression changes with time or with other factors. For instance, the common baker's yeast, *Saccharomyces cerevisiae*, contains about 7000 genes; with a microarray, one can measure qualitatively how each gene is expressed, and how that expression changes, for example, with a change in temperature. There are many different ways to fabricate microarrays; the most common are silicon chips, microscope slides with spots of ~ 100 micrometre diameter, custom arrays, and arrays with larger spots on porous membranes (macroarrays). There can be anywhere from 100 spots to more than 10,000 on a given array.

Arrays can also be made with molecules other than DNA. For example, an antibody array can be used to determine what proteins or bacteria are present in a blood sample.

Allele Specific Oligonucleotide

Allele specific oligonucleotide (ASO) is a technique that allows detection of single base mutations without the need for PCR or gel electrophoresis. Short (20-25 nucleotides in length), labeled probes are exposed to the non-fragmented target DNA. Hybridization occurs with high specificity due to the short length of the probes and even a single base change will hinder hybridization. The target DNA is then washed and the labeled probes that didn't hybridize are removed. The target DNA is then analysed for the presence of the probe via radioactivity or fluorescence. In this experiment, as in most molecular biology techniques, a control must be used to ensure successful experimentation. The Illumina Methylation Assay is an example of a method that takes advantage of the ASO technique to measure one base pair differences in sequence.

Antiquated Technologies

In molecular biology, procedures and technologies are continually being developed and older technologies abandoned. For example, before the advent of DNA gel electrophoresis (agarose or polyacrylamide), the size of DNA molecules was typically determined by rate sedimentation in sucrose gradients, a slow and labor-intensive technique requiring expensive instrumentation; prior to sucrose gradients, viscometry was used.

Aside from their historical interest, it is often worth knowing about older technology, as it is occasionally useful to solve another new problem for which the newer technique is inappropriate.

History

While molecular biology was established in the 1930s, the term was first coined by Warren Weaver in 1938. Warren was the director of Natural Sciences for the Rockefeller Foundation at the time and believed that biology was about to undergo a period of significant change given recent advances in fields such as X-ray crystallography. He therefore channeled significant amounts of (Rockefeller Institute) money into biological fields.

The history of molecular biology begins in the 1930s with the convergence of various, previously distinct biological disciplines: biochemistry, genetics, microbiology, and virology. With the hope of understanding life at its most fundamental level, numerous physicists and chemists also took an interest in what would become molecular biology.

In its modern sense, molecular biology attempts to explain the phenomena of life starting from the macromolecular properties that generate them. Two categories of macromolecules in particular are the focus of the molecular biologist: 1) nucleic acids, among which the most famous is deoxyribonucleic acid (or DNA), the constituent of genes, and 2) proteins, which are the active agents of living organisms. One definition of the scope of molecular biology therefore is to characterize the structure, function and relationships between these two types of macromolecules. This relatively limited definition will suffice to allow us to establish a date for the so-called "molecular revolution", or at least to establish a chronology of its most fundamental developments.

General Overview

In its earliest manifestations, molecular biology—the name was coined by Warren Weaver of the Rockefeller Foundation in 1938—was an ideal of physical and chemical explanations of life, rather than a coherent discipline. Following the advent of the Mendelian-chromosome theory of heredity in the 1910s and the maturation of atomic theory and quantum mechanics in the 1920s, such explanations seemed within reach.

Weaver and others encouraged (and funded) research at the intersection of biology, chemistry and physics, while prominent physicists such as Niels Bohr and Erwin Schrödinger turned their attention to biological speculation. However, in the 1930s and 1940s it was by no means clear which—if any—cross-disciplinary research would bear fruit; work in colloid chemistry, biophysics and radiation biology, crystallography, and other emerging fields all seemed promising.

In 1940, George Beadle and Edward Tatum demonstrated the existence of a precise relationship between genes and proteins. In the course of their experiments connecting genetics with biochemistry, they switched from the genetics mainstay *Drosophila* to a more appropriate model organism, the fungus *Neurospora*; the construction and exploitation of new model organisms would become a recurring theme in the development of molecular biology. In 1944, Oswald Avery, working at the Rockefeller Institute of New York, demonstrated that genes are made up of DNA. In 1952, Alfred Hershey and Martha Chase confirmed that the genetic material of the bacteriophage, the virus which infects bacteria, is made up of DNA. In 1953, James Watson and Francis Crick discovered the double helical structure of the DNA molecule. In 1961, Francois Jacob and Jacques Monod hypothesized the existence of an intermediary between DNA and its protein products, which they called messenger RNA. Between 1961 and 1965, the relationship between the information contained in DNA and the structure of proteins was determined: there is a code, the genetic code, which creates a correspondence between the succession of nucleotides in the DNA sequence and a series of amino acids in proteins. At the beginning of the 1960s, Monod and Jacob also demonstrated how certain specific proteins, called regulative proteins, latch onto DNA at the edges of the genes and control the transcription of these genes into messenger RNA; they direct the "expression" of the genes.

The chief discoveries of molecular biology took place in a period of only about twenty-five years. Another fifteen years were required before new and more sophisticated technologies, united today under the name of genetic engineering, would permit the isolation and characterization of genes, in particular those of highly complex organisms.

The Exploration of the Molecular Dominion

If we evaluate the molecular revolution within the context of biological history, it is easy to note that it is the culmination of a long process which began with the first observations through a microscope. The aim of these early researchers was to understand the functioning of living organisms by describing their organization at the microscopic level. From the end of the 18th century, the characterization of the chemical molecules which make up living beings gained increasingly greater attention, along with the birth of physiological chemistry in the 19th century, developed by the German chemist Justus von Liebig and following the birth of biochemistry at the beginning of the 20th, thanks to another German chemist Eduard Buchner.

Between the molecules studied by chemists and the tiny structures visible under the optical microscope, such as the cellular nucleus or the chromosomes, there was an obscure zone, "the world of the ignored dimensions," as it was called by the chemical-physicist Wolfgang Ostwald. This world is populated by colloids, chemical compounds whose structure and properties were not well defined.

The successes of molecular biology derived from the exploration of that unknown world by means of the new technologies developed by chemists and physicists: X-ray diffraction, electron microscopy, ultracentrifugization, and electrophoresis. These studies revealed the structure and function of the macromolecules.

A milestone in that process was the work of Dr. Linus Pauling in 1949, which for the first time linked the specific genetic mutation in patients with sickle cell disease to a demonstrated change in an individual protein, the hemoglobin in the erythrocytes of heterozygous or homozygous individuals.

The Encounter between Biochemistry and Genetics

The development of molecular biology is also the encounter of two disciplines which made considerable progress in the course of the first thirty years of the twentieth century: biochemistry and genetics. The first studies the structure and function of the molecules which make up living things. Between 1900 and 1940, the central processes of metabolism were described: the process of digestion and the absorption of the nutritive elements derived from alimentation, such as the sugars. Every one of these processes is catalyzed by a particular enzyme. Enzymes are proteins, like the antibodies present in blood or the proteins responsible for muscular contraction. As a consequence, the study of proteins, of their structure and synthesis, became one of the principal objectives of biochemists.

The second discipline of biology which developed at the beginning of the 20th century is genetics. After the rediscovery of the laws of Mendel through the studies of Hugo de Vries, Carl Correns and Erich von Tschermak in 1900, this science began to take shape thanks to the adoption by Thomas Hunt Morgan, in 1910, of a model organism for genetic studies, the famous fruit fly (Drosophila melanogaster). Shortly after, Morgan showed that the genes are localized on chromosomes. Following this discovery, he continued working with Drosophila and, along with numerous other research groups, confirmed the importance of the gene in the life and development of organisms. Nevertheless, the chemical nature of genes and their mechanisms of action remained a mystery. Molecular biologists committed themselves to the determination of the structure, and the description of the complex relations between, genes and proteins.

The development of molecular biology was not just the fruit of some sort of intrinsic "necessity" in the history of ideas, but was a characteristically historical phenomenon, with all of its unknowns, imponderables and contingencies: the remarkable developments in physics at the beginning of the 20th century highlighted the relative lateness in development in biology, which became the "new frontier" in the search for knowledge about the empirical world. Moreover, the developments of the theory of information and cybernetics in the 1940s, in response to military exigencies, brought to the new biology a significant number of fertile ideas and, especially, metaphors.

The choice of bacteria and of its virus, the bacteriophage, as models for the study of the fundamental mechanisms of life was almost natural - they are the smallest living organisms known to exist - and at the same time the fruit of individual choices. This model owes its success, above all, to the fame and the sense of organization of Max Delbrück, a German physicist, who was able to create a dynamic research group, based in the United States, whose exclusive scope was the study of the bacteriophage: the *School of the Phage.*

The geographic panorama of the developments of the new biology was conditioned above all by preceding work. The US, where genetics had developed the most rapidly, and the UK, where there was a coexistence of both genetics and biochemical research of highly advanced levels, were in the avant-garde. Germany, the cradle of the revolutions in physics, with the best minds and the most advanced laboratories of genetics in the world, should have had a primary role in the development of molecular biology. But history decided differently: the arrival of the Nazis in 1933 - and, to a less extreme degree, the rigidification of totalitarian measures in fascist Italy - caused the emigration of a large number of Jewish and non-Jewish scientists. The majority of them fled to the US or the UK, providing an extra impulse to the scientific dynamism of those nations. These movements ultimately made molecular biology a truly international science from the very beginnings.

History of DNA Biochemistry

First Isolation of DNA

Working in the 19th century, biochemists initially isolated DNA and RNA (mixed together) from cell nuclei. They were relatively quick to appreciate the polymeric nature of their "nucleic acid" isolates, but realized only later that nucleotides were of two types—one containing ribose and the other deoxyribose. It was this subsequent discovery that led to the identification and naming of DNA as a substance distinct from RNA.

Friedrich Miescher (1844–1895) discovered a substance he called "nuclein" in 1869. Somewhat later, he isolated a pure sample of the material now known as DNA from the sperm of salmon, and in 1889 his pupil, Richard Altmann, named it "nucleic acid". This substance was found to exist only in the chromosomes.

In 1919 Phoebus Levene at the Rockefeller Institute identified the components (the four bases, the sugar and the phosphate chain) and he showed that the components of DNA were linked in the order phosphate-sugar-base. He called each of these units a nucleotide and suggested the DNA molecule consisted of a string of nucleotide units linked together through the phosphate groups, which are the 'backbone' of the molecule. However Levene thought the chain was short and that the bases repeated in the same fixed order. Torbjorn Caspersson and Einar Hammersten showed that DNA was a polymer.

Chromosomes and Inherited Traits

Max Delbrück, Nikolai V. Timofeeff-Ressovsky, and Karl G. Zimmer published results in 1935 suggesting that chromosomes are very large molecules the structure of which can be changed by treatment with X-rays, and that by so changing their structure it was possible to change the heritable characteristics governed by those chromosomes. In 1937 William Astbury produced the first X-ray diffraction patterns from DNA. He was not able to propose the correct structure but the patterns showed that DNA had a regular structure and therefore it might be possible to deduce what this structure was.

In 1943, Oswald Theodore Avery and a team of scientists discovered that traits proper to the "smooth" form of the *Pneumococcus* could be transferred to the "rough" form of the same bacteria merely by making the killed "smooth" (S) form available to the live "rough" (R) form. Quite unexpectedly, the living R *Pneumococcus* bacteria were transformed into a new strain of the S form, and the transferred S characteristics turned out to be heritable. Avery called the medium of transfer of traits the transforming principle; he identified DNA as the transforming principle, and not protein as previously thought. He essentially redid Frederick Griffith's experiment. In 1953, Alfred Hershey and Martha Chase did an experiment (Hershey-Chase experiment) that showed, in T2 phage, that DNA is the genetic material (Hershey shared the Nobel prize with Luria).

Discovery of the Structure of DNA

In the 1950s, three groups made it their goal to determine the structure of DNA. The first group to start was at King's College London and was led by Maurice Wilkins and was later joined by

Rosalind Franklin. Another group consisting of Francis Crick and James D. Watson was at Cambridge. A third group was at Caltech and was led by Linus Pauling. Crick and Watson built physical models using metal rods and balls, in which they incorporated the known chemical structures of the nucleotides, as well as the known position of the linkages joining one nucleotide to the next along the polymer. At King's College Maurice Wilkins and Rosalind Franklin examined X-ray diffraction patterns of DNA fibers. Of the three groups, only the London group was able to produce good quality diffraction patterns and thus produce sufficient quantitative data about the structure.

Helix Structure

In 1948 Pauling discovered that many proteins included helical shapes. Pauling had deduced this structure from X-ray patterns and from attempts to physically model the structures. (Pauling was also later to suggest an incorrect three chain helical DNA structure based on Astbury's data.) Even in the initial diffraction data from DNA by Maurice Wilkins, it was evident that the structure involved helices. But this insight was only a beginning. There remained the questions of how many strands came together, whether this number was the same for every helix, whether the bases pointed toward the helical axis or away, and ultimately what were the explicit angles and coordinates of all the bonds and atoms. Such questions motivated the modeling efforts of Watson and Crick.

Complementary Nucleotides

In their modeling, Watson and Crick restricted themselves to what they saw as chemically and biologically reasonable. Still, the breadth of possibilities was very wide. A breakthrough occurred in 1952, when Erwin Chargaff visited Cambridge and inspired Crick with a description of experiments Chargaff had published in 1947. Chargaff had observed that the proportions of the four nucleotides vary between one DNA sample and the next, but that for particular pairs of nucleotides — adenine and thymine, guanine and cytosine — the two nucleotides are always present in equal proportions.

Using X-ray diffraction, as well as other data from Rosalind Franklin and her information that the bases were paired, James D. Watson and Francis Crick arrived at the first accurate model of DNA's molecular structure in 1953, which was accepted through inspection by Rosalind Franklin. The discovery was announced on February 28,

1953; the first Watson/Crick paper appeared in Nature on April 25, 1953. Sir Lawrence Bragg, the director of the Cavendish Laboratory, where Watson and Crick worked, gave a talk at Guys Hospital Medical School in London on Thursday, May 14, 1953 which resulted in an article by Ritchie Calder in The News Chronicle of London, on Friday, May 15, 1953, entitled "Why You Are You. Nearer Secret of Life." The news reached readers of The New York Times the next day; Victor K. McElheny, in researching his biography, "Watson and DNA: Making a Scientific Revolution", found a clipping of a six-paragraph New York Times article written from London and dated May 16, 1953 with the headline "Form of `Life Unit' in Cell Is Scanned." The article ran in an early edition and was then pulled to make space for news deemed more important.(The New York Times subsequently ran a longer article on June 12, 1953). The Cambridge University undergraduate newspaper also ran its own short article on the discovery on Saturday, May 30, 1953. Bragg's original announcement at a Solvay conference on proteins in Belgium on 8 April 1953 went unreported by the press. In 1962 Watson, Crick, and Maurice Wilkins jointly received the Nobel Prize for Physiology or Medicine for their determination of the structure of DNA.

"Central Dogma"

Watson and Crick's model attracted great interest immediately upon its presentation. Arriving at their conclusion on February 21, 1953, Watson and Crick made their first announcement on February 28. In an influential presentation in 1957, Crick laid out the "Central Dogma", which foretold the relationship between DNA, RNA, and proteins, and articulated the "sequence hypothesis." A critical confirmation of the replication mechanism that was implied by the double-helical structure followed in 1958 in the form of the Meselson-Stahl experiment. Work by Crick and coworkers showed that the genetic code was based on non-overlapping triplets of bases, called codons, and Har Gobind Khorana and others deciphered the genetic code not long afterward (1966). These findings represent the birth of molecular biology.

History of RNA Tertiary Structure

Pre-history: The Helical Structure of RNA

The earliest work in RNA structural biology coincided, more or less, with the work being done on DNA in the early 1950s. In their seminal 1953 paper, Watson and Crick suggested that van der Waals crowding by the 2'OH group of ribose would preclude RNA from adopting a double helical structure identical to the model they proposed - what we now know as B-form DNA. This provoked questions about the three dimensional structure of RNA: could this molecule form some type of helical structure, and if so, how? As with DNA, early structural work on RNA centered around isolation of native RNA polymers for fiber diffraction analysis.

In part because of heterogeneity of the samples tested, early fiber diffraction patterns were usually ambiguous and not readily interpretable. In 1955, Grunberg-Manago *et al.* published a paper describing the enzyme polynucleotide phosphorylase, which cleaved a phosphate group from nucleotide diphosphates to catalyze their polymerization. This discovery allowed researchers to synthesize homogenous nucleotide polymers, which they then combined to produce double stranded molecules. These samples yielded the most readily interpretable fiber diffraction patterns yet obtained, suggesting an ordered, helical structure for cognate, double stranded RNA that differed from that observed in DNA. These results paved the way for a series of investigations into the various properties and propensities of RNA. Through the late 1950s and early 1960s, numerous papers were publihed on various topics in RNA structure, including RNA-DNA hybridization, triple stranded RNA, and even small-scale crystallography of RNA di-nucleotides - G-C, and A-U - in primitive helix-like arrangements.

The Beginning: Crystal Structure of tRNA[PHE]

In the mid-1960s, the role of tRNA in protein synthesis was being intensively studied. At this point, ribosomes had been implicated in protein synthesis, and it had been shown that an mRNA strand was necessary for the formation of these structures. In a 1964 publication, Warner and Rich showed that ribosomes active in protein synthesis contained tRNA molecules bound at the A and P sites, and discussed the notion that these molecules aided in the peptidyl transferase reaction. However, despite considerable biochemical characterization, the structural basis of tRNA function remained a mystery. In 1965, Holley *et al.* purified and sequenced the first tRNA molecule, initially proposing that it adopted a cloverleaf structure, based largely on the ability of certain regions of the molecule to form stem loop structures. The isolation of tRNA proved to be the first major windfall in RNA structural biology. Following Holley's publication, numerous investigators began work on isolation tRNA for crystallographic study, developing improved methods for isolating the molecule as they worked. By 1968 several groups had produced tRNA crystals, but these proved to be of limited quality and did not yield data at the resolutions necessary to determine structure. In 1971, Kim *et al.* achieved another breakthrough, producing crystals of yeast tRNA[PHE] that diffracted to 2-3 Ångström resolutions by using spermine, a naturally occurring polyamine, which bound to and stabilized the tRNA. Despite having suitable crystals, however, the structure of tRNA[PHE] was not immediately solved at high resolution; rather it took pioneering work in the use of heavy metal derivatives and a good deal more time to produce a high-quality density map of the entire molecule. In 1973, Kim *et al.* produced a 4 Ångström map of the tRNA molecule in which they could unambiguously trace the entire backbone. This solution would be followed by many more, as various investigators worked to refine the structure and thereby more thoroughly elucidate the details of base pairing and stacking interactions, and validate the published architecture of the molecule.

The tRNA[PHE] structure is notable in the field of nucleic acid structure in general, as it represented the first solution of a long-chain nucleic acid structure of any kind - RNA or DNA - preceding Dickerson's solution of a B-form dodecamer by nearly a decade. Also, tRNA[PHE] demonstrated many of the tertiary interactions observed in RNA architecture which would not be categorized and more thoroughly understood for years to come, providing a foundation for all future RNA structural research.

The Renaissance: The Hammerhead Ribozyme and the Group I Intron: P_{4-6}

For a considerable time following the first tRNA structures, the field of RNA structure did not dramatically advance. The ability to study an RNA strcuture depended upon the potential to isolate the RNA target. This proved limiting to the field for many years, in part owing to the fact that other know targets - i.e. the ribosome - were significantly more difficult to isolate and crystallize. Further, because other interesting RNA targets had simply not been identified, or were not sufficiently understood to be deemed interesting, there was simply a lack of things to study structurally. As such, for some twenty years following the original publication of the tRNA[PHE] structure, the structures of only a handful of other RNA targets were solved, with almost all of these belonging to the transfer RNA family. This unfortunate lack of scope would eventually be overcome largely because of two major advancements in nucleic acid research: the identification of ribozymes, and the ability to produce them via *in vitro* transcription.

Subsequent to Tom Cech's publication implicating the *Tetrahymena* group I intron as an autocatalytic ribozyme, and Sidney Altman's report of catalysis by ribonuclease P RNA, several other catalytic RNAs were identified in the late 1980s, including the hammerhead ribozyme. In 1994, McKay *et al.* published the structure of a 'hammerhead RNA-DNA ribozyme-inhibitor complex' at 2.6 Ångström resolution, in which the autocatalytic activity of the ribozyme was disrupted via binding to a DNA substrate. The conformation of the ribozyme published in this paper was eventually shown to be one of several possible states, and although this particular sample was catalytically inactive, subsequent structures have revealed its active-state architecture. This structure was followed by Doudna's publication of the structure of the P4-P6 domains of the *Tetrahymena* group I intron, a fragment of the ribozyme originally made famous by Cech. The second clause in the title of this publication - *Principles of RNA Packing* - concisely evinces the value of these two structures: for the first time, comparisons could be made between well described tRNA structures and those of globular RNAs outside the transfer family. This allowed the framework of categorization to be built for RNA tertiary structure. It was now possible to propose the conservation of motifs, folds, and various local stabilizing interactions.

In addition to the advances being made in global structure determination via crystallography, the early 1990s also saw the implementation of NMR as a powerful technique in RNA structural biology. Coincident with the large-scale ribozyme structures being solved crystallographically, a number of structures of small RNAs and RNAs complexed with drugs and peptides were solved using NMR. In addition, NMR was now being used to investigate and supplement crystal structures, as exemplified by the determination of an isolated tetraloop-receptor motif structure published in 1997. Investigations such as this enabled a more precise characterization of the base pairing and base stacking interactions which stabilized the global folds of large RNA molecules. The importance of understanding RNA tertiary structural motifs was prophetically well described by Michel and Costa in their publication identifying the tetraloop motif: "..it should not come as a surprise if self-folding RNA molecules were to make intensive use of only a relatively small set of tertiary motifs. Identifying these motifs would greatly aid modeling enterprises, which will remain essential as long as the crystallization of large RNAs remains a difficult task".

The Modern Era: The Age of RNA Structural Biology

The resurgence of RNA structural biology in the mid-1990s has caused a veritable explosion in the field of nucleic acid structural research. Since the publication of the hammerhead and P_{4-6} structures, numerous major contributions to the field have been made. Some of the most noteworthy examples include the structures of the Group I and Group II introns, and the Ribosome. It should be noted that the first three structures were produced using *in vitro* transcription, and that NMR has played a role in investigating partial components of all four structures - testaments to the indispensability of both techniques for RNA research. Most recently, the 2009 Nobel Prize in Chemistry was awarded to Ada Yonath, Venkatraman Ramakrishnan, and Thomas Steitz for their structural work on the ribosome, demonstrating the prominent role RNA structural biology has taken in modern molecular biology.

History of Protein Biochemistry

Purifications and Measurements of Mass

The minimum molecular weight suggested by Mulder's analyses was roughly 9 kDa, hundreds of times larger than other molecules being studied. Hence, the chemical structure of proteins (their primary structure) was an active area of research until 1949, when Fred Sanger sequenced insulin. The (correct) theory that proteins were linear polymers of amino acids linked by peptide bonds was proposed independently and simultaneously by Franz Hofmeister and Emil Fischer at the same conference in 1902. However, some scientists were sceptical that such long macromolecules could be stable in solution. Consequently, numerous alternative theories of the protein primary structure were proposed, e.g., the colloidal hypothesis that proteins were assemblies of small molecules, the cyclol hypothesis of Dorothy Wrinch, the diketopiperazine hypothesis of Emil Abderhalden and the pyrrol/piperidine hypothesis of Troensgard (1942). Most of these theories had difficulties in accounting for the fact that the digestion of proteins yielded peptides and amino acids. Proteins were finally shown to be macromolecules of well-defined composition (and not colloidal mixtures) by Theodor Svedberg using analytical ultracentrifugation. The possibility that some proteins are non-covalent associations of such macromolecules was shown by Gilbert Smithson Adair (by measuring the osmotic pressure of hemoglobin) and, later, by Frederic M. Richards in his studies of ribonuclease S. The mass spectrometry of proteins has long been a useful technique for identifying posttranslational modifications and, more recently, for probing protein structure. Most proteins are difficult to purify in more than milligram quantities, even using the most modern methods. Hence, early studies focused on proteins that could be purified in large quantities, e.g., those of blood, egg white, various toxins, and digestive/metabolic enzymes obtained from slaughterhouses. Many techniques of protein purification were developed during World War II in a project led by Edwin Joseph Cohn to purify blood proteins to help keep soldiers alive. In the late 1950s, the Armour Hot Dog Co. purified 1 kg (= one million milligrams) of pure bovine pancreatic ribonuclease A and made it freely available to scientists around the world. This generous act made RNase A the main protein for basic research for the next few decades, resulting in several Nobel Prizes.

Protein Folding and First Structural Models

The study of protein folding began in 1910 with a famous paper by Henrietta Chick and C. J. Martin, in which they showed that the flocculation of a protein was composed of two distinct processes: the precipitation of a protein from solution was *preceded* by another process called denaturation, in which the protein became much less soluble, lost its enzymatic activity and became more chemically reactive. In the mid-1920s, Tim Anson and Alfred Mirsky proposed that denaturation was a reversible process, a correct hypothesis that was initially lampooned by some scientists as "unboiling the egg". Anson also suggested that denaturation was a two-state ("all-or-none") process, in which one fundamental molecular transition resulted in the drastic changes in solubility, enzymatic activity and chemical reactivity; he further noted that the free energy changes upon denaturation were much smaller than those typically involved in chemical reactions. In 1929, Hsien Wu hypothesized that denaturation was protein unfolding, a purely conformational change that resulted in the exposure of amino acid side chains to the solvent. According to this (correct) hypothesis, exposure of aliphatic and reactive side chains to solvent rendered the protein less soluble and more reactive, whereas the loss of a specific conformation caused the loss of enzymatic activity. Although considered plausible, Wu's hypothesis was not immediately accepted, since so little was known of protein structure and enzymology and other factors could account for the changes in solubility, enzymatic activity and chemical reactivity. In the early 1960s, Chris Anfinsen showed that the folding of ribonuclease A was fully reversible with no external cofactors needed, verifying the "thermodynamic hypothesis" of protein folding that the folded state represents the global minimum of free energy for the protein.

The hypothesis of protein folding was followed by research into the physical interactions that stabilize folded protein structures. The crucial role of hydrophobic interactions was hypothesized by Dorothy Wrinch and Irving Langmuir, as a mechanism that might stabilize her cyclol structures. Although supported by J. D. Bernal and others, this (correct) hypothesis was rejected along with the cyclol hypothesis, which was disproven in the 1930s by Linus Pauling (among others). Instead, Pauling championed the idea that protein structure was stabilized mainly by hydrogen bonds, an idea advanced initially by William Astbury (1933). Remarkably, Pauling's incorrect theory about H-bonds resulted in his *correct* models for the secondary structure elements of proteins, the alpha helix and the beta sheet. The hydrophobic interaction was restored to its correct prominence by a famous article in 1959 by Walter Kauzmann on denaturation, based partly on work by Kaj Linderstrøm-Lang. The ionic nature of proteins was demonstrated by Bjerrum, Weber and Arne Tiselius, but Linderstrom-Lang showed that the charges were generally accessible to solvent and not bound to each other (1949).

The secondary and low-resolution tertiary structure of globular proteins was investigated initially by hydrodynamic methods, such as analytical ultracentrifugation and flow birefringence. Spectroscopic methods to probe protein structure (such as circular dichroism, fluorescence, near-ultraviolet and infrared absorbance) were developed in the 1950s. The first atomic-resolution structures of proteins were solved by X-ray crystallography in the 1960s and by NMR in the 1980s. As of 2006, the Protein Data Bank has nearly 40,000 atomic-resolution structures of proteins. In more recent times, cryo-electron microscopy of large macromolecular assemblies and computational protein structure prediction of small protein domains are two methods approaching atomic resolution.

┌─────────┐
│   **2**   │
└─────────┘

Techniques in Molecular Biology

Molecular biology is the branch of science which deals with the structure and function of macromolecules. It has various techniques which are also applied in other branches of biology.

Here are several techniques used in molecular biology which are also related to fields like genetics, recombinant DNA technology etc. These techniques include following:

Expression cloning - One of the most widely used techniques of molecular biology to study protein function is expression cloning. In this technique DNA coding for a protein of interest is cloned. This is done by using PCR and/or restriction enzymes. The DNA encoding the required protein is inserted into a plasmid which is known as an expression vector. This plasmid may include special promoter elements to drive production of the protein of interest, and may also have antibiotic resistance markers to help follow the plasmid.

Polymerase Chain Reaction - is most frequently uses technique for copying DNA into millions of clones within few minutes. It can be used to insert restriction enzymes sites in a DNA fragment or to mutate a particular base sequence and also for determination about the presence of a particular DNA fragment in cDNA libraries.

Gel electrophoresis - is used to separate fragments of DNA, RNA and proteins using different properties like according to size, ionic bond etc. For example agarose gel is used to separate RNA and DNA according to their size and SDS-PAGE is used to separate different proteins.

Blotting techniques - is the technique of probing DNA(Southern blotting), RNA(Northern blotting), or proteins(Western blotting) molecules. Eastern blotting is used to recognize post translational modifications of proteins. Other well known and essential techniques in molecular biology include arrays like DNA microarray, allele specific oligonucleotide(ASO) etc.

Role of Cellular Biology

Main tool used in cellular biology is microscope. A wide range of microscopes including Light microscope, Electron microscope, Fluorescent microscope are used for this purpose. The main purpose of cellular biology is to study different cellular organelles present in all types(prokaryotic-eukaryotic, plant-animal etc) of cells. These organelles include- nucleus, mitochondria, chloroplast, cytoplasm, endoplasmic reticulam, golgi apparatus, cell wall. cell membrane and many more cellular organelles.

Techniques involved in cellular biology other than microscopy include following:

1. Cell culture - in vitro technique of growing cells in laboratory.
2. Immunostaining - is the technique used to locate proteins present in the cell or parts of tissues.
3. Transfection - involve introduction of new gene into a cell
4. Cell fractionation - used to obtain cellular organelles for their study.

Other techniques used in cellular biology include centrifugation, flow cytometry, immunoprecipitation, gene knockdown etc.

Polymerase Chain Reaction

Polymerase chain reaction is a technique of molecular biology which makes multiple copies of DNA from a single fragment. The process of polymerase chain reaction takes place as follows:

1. First the DNA molecule is denatured by heating under the temperature of 94 to 98 degree centigrade.
2. When the strands are separated, they are bound with DNA primers with the temperature 50 to 65 degree Celsius.
3. After annealing, the elongation of the separated DNA strands take place by using heat stable DNA polymerases like Taq (Thermus aquaticus).

There is another method of PCR named reverse transcription PCR which is used to amplify the RNA molecules, while real time PCR is used to measure the DNA o RNA molecule quantitatively.

Expression Cloning

Cloning is the most useful technique of molecular biology because it is used to study the function of the proteins. This technique works in the way that it clones the DNA coding which makes protein, using the process of polymerase chain reaction or restriction enzymes and made multiple copies of that coding. The cloning converts the DNA samples into plasmid vectors. The plasmid either has promoter elements which make possible the production of protein of interest or they have the antibiotic resistant markers.

The plasmid produced from cloning can be inserted in the bacterial cell or the animal cell depending on the mode of experiment. If it is inserted in the bacterial cell, then three processes are involved which make possible the insertion of plasmid into the bacterial cell. These processes are transformation, conjugation and transduction. When the insertion of the cloned plasmid in the animal cell is concerned then the process of transfection is involved. Transfection techniques are of various types for example, microinjection, calcium phosphate transfection and liposome transfection. But it is also possible that the plasmid can be inserted in the animal cell by bacteria or virus. If bacteria are involved for insertion then it is called as bactofention and mainly the specie of Agrobacterium tumafeciens is used for this purpose.

Plasmid can stay in the eukaryotic cell in two forms; either it mixes up with the genome of the host cell which results in the stable transfection or it stays independently in the cell by transient transfection. In both the cases, the protein of interest is in the cell and can be expressed. If the cell contains specific cell signaling factors or inducible promoters then protein is expressed efficiently than without these factors.

Gel Electrophoresis

Gel electrophoresis is the basic technique of molecular biology which is used to separate DNA, RNA or protein molecules. The method involved is that, the molecules are placed in the immobilized gel and an electric current is passed through it. This current separates the molecules according to the size of the strands. Similarly proteins are also separated in the gel by using an SDS-PAGE gel or it can also be separated according to their size and the charge which they carry.

Allele Specific Oligonucleotide

Allele specific technique is used to detect the single base mutations in the genes. It does not need any gel electrophoresis or polymerase chain reaction. In this technique, short probes are used usually 20 to 25 nucleotides long and they are also labeled. The labeled probes are exposed to the non-fragment target DNA; as a result hybridization takes place due to the short length probes. If there is only a single change in the nucleotide base, then the hybridization will stop. After this step, the target DNA is washed and those probes are removed which did not hybridize. Either radiation or florescence detects the presence of probe on the DNA fragment.

Arrays

This is the technique of molecular biology which uses DNA arrays in the form of spots which are attached to the solid support. This solid support is usually microscopic slide which contains the single stranded DNA oligonucleotide fragment present on each spot. This technique is very feasible because it contains large amount of very short spots with DNA fragments on each slide. The spot is so small that it contains a single DNA fragment. This fragment is complementary to a single DNA sequence. The process starts with the extraction of RNA molecule from a tissue which is converted in the form of cDNA. The DNA fragments present on the spots in the slide are then hybridized with these cDNAs. They can be visualized after hybridization.

This technique is useful in the gene expression data as many arrays can be made with the same position of the DNA fragments on the spots, so that the gene expression of healthy and diseases genes can be compared. For example common yeast consists of 7000 genes. With the microarray, it is possible to observe the expression of each gene qualitatively. It is also possible to view the changes in the expression of these genes due to temperature. There are also other methods of construction of arrays instead of suing DNA fragments. For example an antibody array can be used to determine the presence f proteins or bacteria in the sample of blood.

Protein Folding and Modifications

A polypeptide chain is synthesized on large cellular structures, the ribosomes, by a complex process in which assembly of amino acids in a particular sequence is dictated by messenger RNA (mRNA). The nascent polypeptide chain undergoes folding and, in many cases, chemical modification to generate the final protein. Any polypeptide chain containing n residues could, in principle, fold into 8^n conformations. This value is based on the fact that only eight bond angles are stereochemically allowed in the polypeptide backbone. In general, however, all molecules of any protein species adopt a single conformation, called the *native state,* which is the most stably folded form of the molecule. Misfolding to non-native conformations is suppressed by two distinct mechanisms. At the molecular level, a protein folds through a pathway that favors only a few intermediate steps. Furthermore, a cellular system prevents misfolded proteins from forming. After a protein has carried out its functions, specific sequences that limit the life span of the protein target it for degradation.

The Information for Protein Folding is Encoded in the Sequence

The realization that the amino acid sequence of a protein determines its folding came from in vitro studies on protein unfolding and refolding. Thermal energy from heat, extremes of pH that alter the charges on amino acid side chains, and chemicals such as urea or guanidine hydrochloride at concentrations of 6–8 M can disrupt the weak noncovalent bonds that stabilize the native conformation of a protein. The denaturation resulting from such treatment causes a protein to lose both its compact conformation and activity. Most denatured proteins precipitate in solution because hydrophobic groups, normally buried inside the molecules, interact with similar regions of other unfolded molecules, causing them to form an insoluble aggregate.

Many proteins that are completely unfolded in 8 M urea and β-mercaptoethanol (which reduces disulfide bonds) can *renature* (refold) into their native state when the denaturing reagents are removed by dialysis. During renaturation, all the disulfide, hydrogen, and hydrophobic bonds that stabilize the native conformation are re-formed. Thus, in this case proteins can be carried through a denaturationrenaturation cycle, which first destroys and then reestablishes their original structure and function. Because renaturation requires no cofactors or other proteins, at least in the test tube, protein folding is a self-assembly process.

Treatment with an 8 M urea solution containing mercaptoethanol ($HSCH_2CH_2OH$) completely denatures most proteins. The urea breaks intramolecular hydrogen and hydrophobic bonds, and the mercaptoethanol reduces each disulfide bridge (–S–S–) to two sulfhydryl (–SH) groups. When these chemicals are removed by dialysis, the –SH groups on the unfolded chain oxidize spontaneously to reform disulfide bridges, and the polypeptide chain simultaneously refolds into its native conformation. The observation by Christian Anfinsen of such reversible denaturation and renaturation of ribonuclease, an enzyme that degrades RNA, provided a clue that the information for folding a protein lies in its sequence. A general mechanism by which proteins refold in vitro has been elucidated in experiments in which the renaturing conditions are carefully adjusted and the refolding reaction is interrupted at various time intervals. Such studies have shown that the polypeptide goes through several transient reconfigurations, including a "molten globule" state, before the native tertiary conformation is reached. In the case of ribonuclease, which has several internal disulfide bonds, the folding pathway involves rearrangements of disulfide bond pairs to the native conformation.

Under appropriate refolding conditions, the molecule condenses around a hydrophobic core into a compact, but non-native, intermediate, called a *molten globule.*In this folding intermediate, much of the secondary structure is present. Long-range interactions then form the tertiary structure, folding the molecule into its native three-dimensional conformation.

Folding of Proteins in Vivo is Promoted by Chaperones

Folding of proteins in vitro is an inefficient process, with only a minority of unfolded molecules undergoing complete folding within a few minutes. Clearly, in vivo most protein molecules must rapidly fold into their correct shape; otherwise, cells would waste much energy in the synthesis of nonfunctional proteins and in the degradation of misfolded or unfolded proteins. More than 95 percent of the proteins present within cells have been shown to be in their native conformation, despite high protein concentrations (H–100 mg/ml), which usually cause proteins to precipitate in vitro.

The explanation for the cell's remarkable efficiency in promoting protein folding probably lies in chaperones, a family of proteins found in all organisms from bacteria to humans. Chaperones are located in every cellular compartment, bind a wide range of proteins, and may be part of a general protein-folding mechanism. There are two general families of chaperones: *molecular chaperones,* which bind and stabilize unfolded or partially folded proteins, thereby preventing these proteins from being degraded; and *chaperonins,* which directly facilitate their folding. Chaperones have ATPase activity, and their ability to bind and stabilize their target proteins is specific and dependent on ATP hydrolysis. Binding of chaperones to partially folded proteins suggests that the folding process could be regulated at intermediate steps.

Molecular chaperones consist of the Hsp70 family of proteins, which includes Hsp70 in the cytosol and mitochondrial matrix, Bip in the endoplasmic reticulum, and DnaK, a bacterial chaperone. First identified by its rapid appearance after a cell has been stressed by heat shock, Hsp70 is the major chaperone protein in all organisms. When bound to ATP, Hsp70 assumes an open form in which an exposed hydrophobic pocket transiently binds to exposed hydrophobic regions of the unfolded target protein. Hydrolysis of the bound ATP causes Hsp70 to assume a closed form, releasing the target protein. Molecular chaperones are thought to bind all nascent polypeptide chains as they are being synthesized on ribosomes. In bacteria, 85 percent of the proteins are released from their chaperone and go on to fold normally; an even higher percentage of proteins in eukaryotes follow this pathway.

(a) Many proteins 1 fold into their proper three-dimensional structure with the assistance of Hsp70, a molecular chaperone that transiently binds to a nascent polypeptide as it emerges from a ribosome. Proper folding of some proteins 2 also depends on the chaperonin TCiP, a large barrel-shaped complex of Hsp60 units.

(b) GroEL, the bacterial homolog of TCiP, is a barrel-shaped complex of 14 identical 60,000-MW subunits arranged in two stacked rings. In the absence of ATP or presence of ADP, GroEL exists in a "tight" conformational state that binds partially folded or misfolded proteins. Binding of ATP shifts GroEL to a more open, "relaxed" state, which releases the folded protein.

Proper folding of a small proportion of proteins (e.g., the cytoskeletal proteins actin and tubulin) requires additional assistance, which is provided by chaperonins. Eukaryotic chaperonins, called *TCiP*, are large, barrel-shaped, multimeric complexes composed of eight Hsp60 units. The bacterial homolog, known as *GroEL*, contains 14 identical subunits. The GroEL folding mechanism, which is better understood than TCiP-mediated folding, serves as a reasonable general model. In bacteria, a partially folded or misfolded polypeptide is inserted into the cavity of GroEL, where it binds to the inner wall and folds into its native conformation.

In an ATP-dependent step, GroEL expands and the protein exits GroEL, a process assisted by a co-chaperonin, GroES, which caps the ends of GroEL. Because the eukaryotic chaperonin TCiP lacks a GroES-type co-chaperonin, the last step must differ in eukaryotes. Moreover, the size of the cavity in TCiP limits this folding pathway to polypeptides smaller than 55 kDa.

Chemical Modifications and Processing Alter the Biological Activity of Proteins

Nearly every protein in a cell is chemically altered after its synthesis on a ribosome. Such modifications may alter the activity, life span, or cellular location of proteins, depending on the nature of the alteration. Protein alterations fall into two categories: chemical modification and processing. *Chemical modification* involves the linkage of a chemical group to the terminal amino or carboxyl groups or to reactive groups in the side chains of internal residues; in some cases, these modifications are reversible. *Processing* involves the removal of peptide segments and generally is irreversible.

Acetylation, the addition of an acetyl group (CH_3CO) to the amino group of the N-terminal residue is the most common form of chemical modification, involving an estimated 80 percent of all proteins:

This modification may play an important role in controlling the life span of proteins within cells, as nonacetylated proteins are rapidly degraded by intracellular proteases. As discussed later, residues at or near the termini of some membrane proteins are chemically modified by addition of long lipidlike groups. Attachment of these hydrophobic "tails," which function to anchor proteins to the lipid bilayer, constitute one way that cells restrict certain proteins to membranes.

The internal residues in proteins can be modified by attachment of a variety of chemical groups to their side chains. The most important modification is *phosphorylation* of serine, threonine, and tyrosine residues. We will encounter numerous examples of proteins whose activity is regulated by reversible phosphorylation and dephosphorylation. The side chains of asparagine, serine, and threonine are sites for *glycosylation,* the attachment of linear and branched carbohydrate chains. Many secreted proteins and membrane proteins contain glycosylated residues. Various less common modifications are found in a limited number of proteins.

Examples of modified internal residues produced by hydroxylation, methylation, and carboxylation. These modifications occur after synthesis of the polypeptide chain.

These modifications occur after synthesis of the polypeptide chain.

Unlike chemical modification of residues, which often is reversible, processing of some proteins causes irreversible changes that alter their activity. In the most common form of processing, residues are removed from the C- or N-terminus of a polypeptide by cleavage of the peptide bond in a reaction catalyzed by proteases. Proteolytic cleavage is a common mechanism of activation or inactivation, especially of enzymes involved in blood coagulation or digestion. As discussed later, the activity of certain digestive enzymes is controlled by this mechanism. Proteolysis also generates active peptide hormones, such as EGF mentioned earlier and insulin, from larger precursor polypeptides.

An unusual type of processing, termed *protein selfsplicing*, occurs in bacteria and primitive eukaryotes. *Splicing* refers to a process analogous to editing film: an internal segment of polypeptide, an intein, is removed and the ends of the polypeptide are rejoined. Unlike proteolytic processing, protein self-splicing is an autocatalytic process, which proceeds by itself without the involvement of enzymes. The excised peptide appears to eliminate itself from the protein by a mechanism similar to that used in processing of RNA molecules. In vertebrate cells, processing of some proteins involves self-cleavage, but the subsequent ligation step is absent. One such protein is Hedgehog, which is critical to a number of developmental processes.

A segment (red) of a polypeptide, called an *intein,* is removed by cleavage of two peptide bonds (red arrows), leaving two segments (blue and green). One new peptide bond then forms between the two segments (blue arrow), regenerating a continuous polypeptide backbone. The process is autocatalytic and does not depend on enzymes. Consensus splice sites in the polypeptide chain mark the points of breakage and re-formation of the backbone. The excised segment is thought to be exposed at the surface of the folded protein.

Cells Degrade Proteins Via Several Pathways

Cells have both extracellular and intracellular pathways for degrading proteins. The major extracellular pathway is the system of *digestive proteases,* which break down ingested proteins to polypeptides in the intestinal tract. These include *endoproteases* such as trypsin and chymotrypsin, which cleave the protein backbone adjacent to basic and aromatic residues; *exopeptidases,* which sequentially remove residues from the N-terminus (aminopeptidases) or C-terminus (carboxypeptidases) of proteins; and *peptidases,* which split oligopeptides into di- and tripeptides and individual amino acids. These small molecules then are transported across the intestinal lining into the bloodstream.

The life span of intracellular proteins varies from as short as a few minutes for mitotic cyclins, which help regulate passage through mitosis, to as long as the age of an organism for proteins in the lens of the eye. Cells have several intracellular proteolytic pathways for degrading misfolded or denatured proteins, normal proteins whose concentration must be decreased, and foreign proteins taken up by the cell. One major intracellular pathway involves degradation by enzymes within lysosomes, membrane-limited organelles whose interior is acidic. Distinct from the lysosomal pathway are cytosolic mechanisms for degrading proteins. The best-understood pathway, the ubiquitin-mediated pathway, involves two steps: addition of a chain of ubiquitin molecules to an internal lysine side chain of a target protein and proteolysis of the ubiquitinated protein by a proteasome, a large, cylindrical multisubunit complex. The numerous proteasomes present in the cell cytosol proteolytically cleave ubiquitin-tagged proteins in an ATP-dependent process that yields peptides and intact ubiquitin molecules.

A conjugating enzyme catalyzes formation of a peptide bond between ubiquitin (Ub) and the side-chain $-NH_2$ of a lysine residue in a target protein. Additional Ub molecules are added, forming a multiubiquitin chain. This chain is thought to direct the tagged protein to a proteasome, which cleaves the protein into numerous small peptide fragments. (b) Computer-generated image reveals cylindrical structure of a proteasome with a cap at each end of a core. Proteolysis of ubiquitin-tagged proteins occurs along the inner wall of the core.

To be targeted for degradation by the ubiquitinmediated pathway, a protein must contain a structure that is recognized by a ubiquitinating enzyme complex. Different conjugating enzymes recognize different degradation signals in target proteins. For example, the internal sequence Arg-X-X-Leu-Gly-X-Ile-Gly-Asx in mitotic cyclin is recognized by the ubiquitin-conjugating enzyme E1. Internal sequences enriched in proline, glutamic acid, serine, and threonine (PEST sequences) are recognized by other enzymes. The life span of many cytosolic proteins is correlated with the identity of the N-terminal residue, suggesting that certain residues at the N-terminus favor rapid ubiquitination.

For example, short-lived proteins that are degraded within 3 minutes in vivo commonly have Arg, Lys, Phe, Leu, or Trp at their N-terminus. In contrast, a stabilizing amino acid such as Cys, Ala, Ser, Thr, Gly, Val, or Met is present at the N-terminus in long-lived proteins that resist proteolytic attack for more than 30 hours. All newly synthesized proteins have methionine, a stabilizing amino acid, at the N-terminus. Thus subsequent enzymatic alteration that generates one of the destabilizing amino acids at the N-terminus is necessary to target a protein for degradation.

Aberrantly Folded Proteins are Implicated in Slowly Developing Diseases

As noted earlier, each protein species normally folds into a single, energetically favorable conformation that is specified by its amino acid sequence. Recent evidence suggests, however, that a protein may fold into an alternative three- dimensional structure for reasons that have not yet been identified. Such "misfolding" not only leads to a loss of the normal function of a protein but also marks it for proteolytic degradation. The subsequent accumulation of proteolytic fragments contributes to certain degenerative diseases characterized by the presence of insoluble protein plaques in various organs including the liver and brain.

Alzheimer's disease, for example, is marked by formation of plaques and tangles in a deteriorating brain. The filaments composing these structures are derived from proteolytic products of abundant natural proteins such as amyloid precursor protein, a transmembrane protein, and Tau, a microtubule-binding protein. Plaques in other organs are formed from proteolytic fragments of natural proteins such as gelsolin, an actin-binding protein, and serum albumin, a blood protein. The polypeptide fragments liberated by proteolysis polymerize into very stable filaments. A degeneration of the brain, similar to that seen in Alzheimer's disease, is thought to be caused by *prions,* an infectious protein agent derived by proteolysis and re-folding of a normal brain protein. The amyloid plaque in the brain of an Alzheimer's patient appears as a tangle of filaments. (b) In the atomic force microscope, the filaments are seen to be regular arrangements of a short 47-residue fragment, called β-*amyloid peptide,* produced by proteolysis of amyloid precursor protein.

  • The amino acid sequence of a protein dictates its folding into a specific three-dimensional conformation, the native state.

- Folding of denatured proteins in vitro proceeds through intermediates having secondary and non-native tertiary structure.

- Protein folding in vivo occurs with the assistance of two types of special proteins. Molecular chaperones (Hsp70 proteins) bind to nascent polypeptides emerging from ribosomes and prevent their misfolding. Chaperonins, large complexes of Hsp60-like proteins, shelter some partially folded or misfolded proteins in a barrel-like cavity, providing additional time for proper folding.

- Following their synthesis, all proteins are modified in various ways that alter their structure and function.

- The life span of intracellular proteins is largely determined by their susceptibility to proteolytic degradation by various pathways.

- The presence of certain internal sequences or Nterminal residues targets cytosolic proteins for addition of ubiquitin and subsequent proteolysis within a proteasome.

Genomics and Proteomics

Genomics is a discipline in genetics concerning the study of the genomes of organisms. The field includes intensive efforts to determine the entire DNA sequence of organisms and fine-scale genetic mapping efforts. The field also includes studies of intragenomic phenomena such as heterosis, epistasis, pleiotropy and other interactions between loci and alleles within the genome.

In contrast, the investigation of the roles and functions of single genes is a primary focus of molecular biology or genetics and is a common topic of modern medical and biological research. Research of single genes does not fall into the definition of genomics unless the aim of this genetic, pathway, and functional information analysis is to elucidate its effect on, place in, and response to the entire genome's networks.

For the United States Environmental Protection Agency, "the term "genomics" encompasses a broader scope of scientific inquiry associated technologies than when genomics was initially considered. A genome is the sum total of all an individual organism's genes. Thus, genomics is the study of all the genes of a cell, or tissue, at the DNA (genotype), mRNA (transcriptome), or protein (proteome) levels."

History

The first genomes to be sequenced were those of a virus and a mitochondrion, and were done by Fred Sanger. His group established techniques of sequencing, genome mapping, data storage, and bioinformatic analyses in the 1970-1980s. A major branch of genomics is still concerned with sequencing the genomes of various organisms, but the knowledge of full genomes has created the possibility for the field of functional genomics, mainly concerned with patterns of gene expression during various conditions.

The most important tools here are microarrays and bioinformatics. Study of the full set of proteins in a cell type or tissue, and the changes during various conditions, is called proteomics. A related concept is materiomics, which is defined as the study of the material properties of biological materials (e.g. hierarchical protein structures and materials, mineralized biological tissues, etc.) and their effect on the macroscopic function and failure in their biological context, linking processes, structure and properties at multiple scales through a materials science approach. The actual term 'genomics' is thought to have been coined by Dr. Tom Roderick, a geneticist at the Jackson Laboratory (Bar Harbor, ME) over beer at a meeting held in Maryland on the mapping of the human genome in 1986.

In 1972, Walter Fiers and his team at the Laboratory of Molecular Biology of the University of Ghent (Ghent, Belgium) were the first to determine the sequence of a gene: the gene for Bacteriophage MS2 coat protein. In 1976, the team determined the complete nucleotide-sequence of bacteriophage MS2-RNA. The first DNA-based genome to be sequenced in its entirety was that of bacteriophage Φ-X174; (5,368 bp), sequenced by Frederick Sanger in 1977.

The first free-living organism to be sequenced was that of *Haemophilus influenzae* (1.8 Mb) in 1995, and since then genomes are being sequenced at a rapid pace.

As of September 2007, the complete sequence was known of about 1879 viruses, 577 bacterial species and roughly 23 eukaryote organisms, of which about half are fungi. Most of the bacteria whose genomes have been completely sequenced are problematic disease-causing agents, such as *Haemophilus influenzae*. Of the other sequenced species, most were chosen because they were well-studied model organisms or promised to become good models. Yeast (*Saccharomyces cerevisiae*) has long been an important model organism for the eukaryotic cell, while the fruit fly *Drosophila melanogaster* has been a very important tool (notably in early pre-molecular genetics). The worm *Caenorhabditis elegans* is an often used simple model for multicellular organisms. The zebrafish *Brachydanio rerio* is used for many developmental studies on the molecular level and the flower *Arabidopsis thaliana* is a model organism for flowering plants. The Japanese pufferfish (*Takifugu rubripes*) and the spotted green pufferfish (*Tetraodon nigroviridis*) are interesting because of their small and compact genomes, containing very little non-coding DNA compared to most species. The mammals dog (*Canis familiaris*), brown rat (*Rattus norvegicus*), mouse (*Mus musculus*), and chimpanzee (*Pan troglodytes*) are all important model animals in medical research.

Human Genomics

A rough draft of the human genome was completed by the Human Genome Project in early 2001, creating much fanfare. By 2007 the human sequence was declared "finished" (less than one error in 20,000 bases and all chromosomes assembled). Display of the results of the project required significant bioinformatics resources. The sequence of the human reference assembly can be explored using the UCSC Genome Browser.

Bacteriophage Genomics

Bacteriophages have played and continue to play a key role in bacterial genetics and molecular biology. Historically, they were used to define gene structure and gene regulation. Also the first genome to be sequenced was a bacteriophage. However, bacteriophage research did not lead the genomics revolution, which is clearly dominated by bacterial genomics. Only very recently has the study of bacteriophage genomes become prominent, thereby enabling researchers to understand the mechanisms underlying phage evolution. Bacteriophage genome sequences can be obtained through direct sequencing of isolated bacteriophages, but can also be derived as part of microbial genomes. Analysis of bacterial genomes has shown that a substantial amount of microbial DNA consists of prophage sequences and prophage-like elements. A detailed database mining of these sequences offers insights into the role of prophages in shaping the bacterial genome.

Cyanobacteria Genomics

At present there are 24 cyanobacteria for which a total genome sequence is available. 15 of these cyanobacteria come from the marine environment. These are six *Prochlorococcus* strains, seven marine *Synechococcus* strains, *Trichodesmium erythraeum* IMS101 and *Crocosphaera watsonii* WH8501. Several studies have demonstrated how these sequences could be used very successfully to infer important ecological and physiological characteristics of marine cyanobacteria.

However, there are many more genome projects currently in progress, amongst those there are further *Prochlorococcus* and marine *Synechococcus* isolates, *Acaryochloris* and *Prochloron*, the N_2-fixing filamentous cyanobacteria *Nodularia spumigena*, *Lyngbya aestuarii* and *Lyngbya majuscula*, as well as bacteriophages infecting marine cyanobaceria.

Thus, the growing body of genome information can also be tapped in a more general way to address global problems by applying a comparative approach. Some new and exciting examples of progress in this field are the identification of genes for regulatory RNAs, insights into the evolutionary origin of photosynthesis, or estimation of the contribution of horizontal gene transfer to the genomes that have been analysed.

Proteomics

Proteomics is the large-scale study of proteins, particularly their structures and functions. Proteins are vital parts of living organisms, as they are the main components of the physiological metabolic pathways of cells. The term "proteomics" was first coined in 1997 to make an analogy with genomics, the study of the genes.

The word "proteome" is a blend of "protein" and "genome", and was coined by Marc Wilkins in 1994 while working on the concept as a PhD student. The proteome is the entire complement of proteins, including the modifications made to a particular set of proteins, produced by an organism or system. This will vary with time and distinct requirements, or stresses, that a cell or organism undergoes.

Complexity of the Problem

After genomics, proteomics is considered the next step in the study of biological systems. It is much more complicated than genomics mostly because while an organism's genome is more or less constant, the proteome differs from cell to cell and from time to time. This is because distinct genes are expressed in distinct cell types. This means that even the basic set of proteins which are produced in a cell needs to be determined.

In the past this was done by mRNA analysis, but this was found not to correlate with protein content. It is now known that mRNA is not always translated into protein, and the amount of protein produced for a given amount of mRNA depends on the gene it is transcribed from and on the current physiological state of the cell. Proteomics confirms the presence of the protein and provides a direct measure of the quantity present.

Post-translational Modifications

Not only does the translation from mRNA cause differences, many proteins are also subjected to a wide variety of chemical modifications after translation. Many of these post-translational modifications are critical to the protein's function.

Phosphorylation

One such modification is phosphorylation, which happens to many enzymes and structural proteins in the process of cell signaling. The addition of a phosphate to particular amino acids—most commonly serine and threonine mediated by serine/threonine kinases, or more rarely tyrosine mediated by tyrosine kinases—causes a protein to become a target for binding or interacting with a distinct set of other proteins that recognize the phosphorylated domain.

Because protein phosphorylation is one of the most-studied protein modifications, many "proteomic" efforts are geared to determining the set of phosphorylated proteins in a particular cell or tissue-type under particular circumstances. This alerts the scientist to the signaling pathways that may be active in that instance.

Ubiquitination

Ubiquitin is a small protein that can be affixed to certain protein substrates by enzymes called E3 ubiquitin ligases. Determining which proteins are poly-ubiquitinated can be helpful in understanding how protein pathways are regulated. This is therefore an additional legitimate "proteomic" study. Similarly, once it is determined what substrates are ubiquitinated by each ligase, determining the set of ligases expressed in a particular cell type will be helpful.

Additional Modifications

Listing all the protein modifications that might be studied in a "Proteomics" project would require a discussion of most of biochemistry; therefore, a short list will serve here to illustrate the complexity of the problem. In addition to phosphorylation and ubiquitination, proteins can be subjected to (among others) methylation, acetylation, glycosylation, oxidation and nitrosylation. Some proteins undergo ALL of these modifications, often in time-dependent combinations, aptly illustrating the potential complexity one has to deal with when studying protein structure and function.

Distinct Proteins are Made under Distinct Settings

Even if one is studying a particular cell type, that cell may make different sets of proteins at different times, or under different conditions. Furthermore, as mentioned, any one protein can undergo a wide range of post-translational modifications. Therefore a "proteomics" study can become quite complex very quickly, even if the object of the study is very restricted. In more ambitious settings, such as when a biomarker for a tumour is sought - when the proteomics scientist is obliged to study sera samples from multiple cancer patients - the amount of complexity that must be dealt with is as great as in any modern biological project.

Limitations to Genomic Study

Scientists are very interested in proteomics because it gives a much better understanding of an organism than genomics. First, the level of transcription of a gene gives only a rough estimate of its level of expression into a protein. An mRNA produced in abundance may be degraded rapidly or translated inefficiently, resulting in a small amount of protein. Second, as mentioned above many proteins experience post-translational modifications that profoundly affect their activities; for example some proteins are not active until they become phosphorylated. Methods such as phosphoproteomics and glycoproteomics are used to study post-translational modifications. Third, many transcripts give rise to more than one protein, through alternative splicing or alternative post-translational modifications. Fourth, many proteins form complexes with other proteins or RNA molecules, and only function in the presence of these other molecules. Finally, protein degradation rate plays an important role in protein content.

Methods of Studying Proteins

Determining Proteins which are Post-translationally Modified

One way in which a particular protein can be studied is to develop an antibody which is specific to that modification. For example, there are antibodies which only recognize certain proteins when they are tyrosine-phosphorylated, known as phospho-specific antibodies; also, there are antibodies specific to other modifications. These can be used to determine the set of proteins that have undergone the modification of interest.

For sugar modifications, such as glycosylation of proteins, certain lectins have been discovered which bind sugars. These too can be used.

A more common way to determine post-translational modification of interest is to subject a complex mixture of proteins to electrophoresis in "two-dimensions", which simply means that the proteins are electrophoresed first in one direction, and then in another... this allows small differences in a protein to be visualized by separating a modified protein from its unmodified form. This methodology is known as "two-dimensional gel electrophoresis".

Recently, another approach has been developed called PROTOMAP which combines SDS-PAGE with shotgun proteomics to enable detection of changes in gel-migration such as those caused by proteolysis or post translational modification.

Determining the Existence of Proteins in Complex Mixtures

Classically, antibodies to particular proteins or to their modified forms have been used in biochemistry and cell biology studies. These are among the most common tools used by practicing biologists today.

For more quantitative determinations of protein amounts, techniques such as ELISAs can be used.

For proteomic study, more recent techniques such as matrix-assisted laser desorption/ionization (MALDI) have been employed for rapid determination of proteins in particular mixtures and increasingly electrospray ionization (ESI).

Computational Methods in Studying Protein Biomarkers

Computational predictive models have shown that extensive and diverse feto-maternal protein trafficking occurs during pregnancy and can be readily detected non-invasively in maternal whole blood. This computational approach circumvented a major limitation, the abundance of maternal proteins interfering with the detection of fetal proteins, to fetal proteomic analysis of maternal blood. Computational models can use fetal gene transcripts previously identified in maternal whole blood to create a comprehensive proteomic network of the term neonate. Such work shows that the fetal proteins detected in pregnant woman's blood originate from a diverse group of tissues and organs from the developing fetus. The proteomic networks contain many biomarkers that are proxies for development and illustrate the potential clinical application of this technology as a way to monitor normal and abnormal fetal development.

An information theoretic framework has also been introduced for biomarker discovery, integrating biofluid and tissue information. This new approach takes advantage of functional synergy between certain biofluids and tissues with the potential for clinically significant findings not possible if tissues and biofluids were considered individually. By conceptualizing tissue-biofluid as information channels, significant biofluid proxies can be identified and then used for guided development of clinical diagnostics. Candidate biomarkers are then predicted based on information transfer criteria across the tissue-biofluid channels. Significant biofluid-tissue relationships can be used to prioritize clinical validation of biomarkers.

Establishing Protein-protein Interactions

Most proteins function in collaboration with other proteins, and one goal of proteomics is to identify which proteins interact. This is especially useful in determining potential partners in cell signaling cascades. Several methods are available to probe protein-protein interactions. The traditional method is yeast two-hybrid analysis. New methods include protein microarrays, immunoaffinity chromatography followed by mass spectrometry, dual polarisation interferometry, Microscale Thermophoresis and experimental methods such as phage display and computational methods

Practical Applications of Proteomics

One of the most promising developments to come from the study of human genes and proteins has been the identification of potential new drugs for the treatment of disease. This relies on genome and proteome information to identify proteins associated with a disease, which computer software can then use as targets for new drugs. For example, if a certain protein is implicated in a disease, its 3D structure provides the information to design drugs to interfere with the action of the protein. A molecule that fits the active site of an enzyme, but cannot be released by the enzyme, will inactivate the enzyme. This is the basis of new drug-discovery tools, which aim to find new drugs to inactivate proteins involved in disease. As genetic differences among individuals are found, researchers expect to use these techniques to develop personalized drugs that are more effective for the individual.

Biomarkers

The FDA defines a biomarker as, "A characteristic that is objectively measured and evaluated as an indicator of normal biologic processes, pathogenic processes, or pharmacologic responses to a therapeutic intervention".

Understanding the proteome, the structure and function of each protein and the complexities of protein-protein interactions will be critical for developing the most effective diagnostic techniques and disease treatments in the future.

An interesting use of proteomics is using specific protein biomarkers to diagnose disease. A number of techniques allow to test for proteins produced during a particular disease, which helps to diagnose the disease quickly. Techniques include western blot, immunohistochemical staining, enzyme linked immunosorbent assay (ELISA) or mass spectrometry.

Current Research Methodologies

There are many approaches to attempting to characterize the human proteome, which is estimated to exceed 100,000 unique forms, 25,000 genes plus post-translational modifications.

In addition, first promising attempts to decipher the proteom of animal tumors have recently been reported.

3

The Chemical Basis of Life: Atoms and Molecules

All matter is made up of tiny particles called *atoms*. Atoms are the basic building blocks of all mater, including life. They cannot be broken down by any known means. There are only about 100 kinds of atoms but they can be chemically combined to make thousands of different kinds of substances. The substances are called *elements* if they are made up of only one kind of atom. Hydrogen gas is an example of an element because there are two atoms of Hydrogen joined together.

Elements are pure sunstances that cannot be broken down by any oridnay means. Unlike solutions, elements are made up of only one kind of atom with the same atomic number. For example, the element Nitrogen consists entirely of Nitrogen atoms. There are over 100 different kinds of elements known today. These elements are organized into different families and different periods in the periodic table. They can be in the form of any state, solid, liquid, or a gas. Compounds are made up of two or more different *elements* put together in definite proportions. For example, water is a compound because it contains two atoms of Hydrogen for every one atom of Oxygen. A molecule is like a compound in that it is two or more atoms combined, but each atom when combined has to act as a single particle. If a sodium atom passes an electron to a chlorine atom then the chlorine atom has 17 protons and 18 electrons. An atom that has an excess charge is called an ion.

Compounds

Most substances are compounds. A compound is made up of two or more different kinds of atoms combined in definite proportions.

Compounds can be separated into the individual atoms that compose their composition, unlike elements that cannot. Compounds can be seperated only by a chemical change that involves heat or electrcity. For example, Water can be separated int Hydrogen and Oxygen by applying a sufficient DC voltage.

Ionic Bonding

Ionic bonding is the bond between two atoms formed as a result of the transfer of electrons from one atom to the other. This results in the formation of positive and negative ions. IONS - Any atom which has an excess charge is called an ion, and the force of attraction between two ions is called an ionic bond. Example: When a sodium atom gives up an electron, it will have 11 protons and 10 electrons. This creates a sodium ion which has an excess positive charge of 1 unit. Relate the concepts of solutions, mixtures, acids and bases to the composition and environment of cells.

A solution is a type of mixture that has the same composition throughtout, and appears to be all one substance. In other words, it's a homogeneous mixture that can be in the form of a solid, liquid or gas. The make up of cells consist of many types of mixtures, including solutions. Many of them are in the form of a liquid. A solution is formed by mixing one or more substances into another substance. In any solution, there is a solute and a solvent. The solvent is generally a liquid, and it is the substance in which the solute is dissolved in. A mixture is a word used to describe a blend of different substances. A mixture is made up of two or more substances, in which they can all be visually seperated and distinguished from one another.

Solutions

A solution is any homogeneous mixture, and may include solids, liquids, as well as gases. Most solutions important to living cells are thought of as liquids. The liquid substance that makes up the bulk of a solution is known as a solvent. Substances that dissolve in the solvent are called solutes (SAHL yoots). A solute can either be a solid, liquid, or gas before they are dissolved in the solvent. The most common solutions have water as the solvent.

Mixtures

A mixture is a combination of substances that are physically mixed without forming new chemical bonds. The substances in a mixture may be present in any proportions. The proportions can change as one substance may be added to, or removed from the mixture, but the different substances in the mixture retain their usual properties.

Acids

An acid is any compound that produces hydrogen ions in a solution. All acids consist of molecules in which contain hydrogen. The molecules are covalently bonded to another atom or group of atoms. When these molecules dissociate in a solution, such as water, the hydrogen breaks loose as an hydrogen ion. The left over molecules form what is called a negative ion. Hydrochloric acid is an example of an acid, which is important to life activities. This is because the level of acid in the environment of the cell affects the actions of enzymes and therefore affects the functioning of the cell.

Base

A base is any compound that produces excess hydroxide ions when it is dissolved in a solution. When in a dry state many bases are ionic compounds. A perfect example of a base is sodium hydroxide. Sodium hydroxide is a solid compound, which is made up of sodium and hydroxide ions. When a base is dissolved in water, it dissociates (separates) into its ions:

$$NaOH \longrightarrow Na^+ + OH^-$$

An acid is any compound that produces hydrogen ions in solution. All acids consist of molecules that contain hydrogen covalently bonded to another atom or group of atoms.

A base is a compound that produces hydroxide(hy DRAHK syd) ions (OH-) when disolved in water. In the dry slate, many bases are ionic compounds. Sodium hydroxide (NaOH) is one example. This is a solid compound made of sodium and hydroxide ions. When dissolved in water it separates in to it's ions. NaOH then dissociates to form Na+ and OH- ions.

When acid and a base react their hydrogen and hydroxide combine to form water molecules. This reaction removes these two ions from the solution. that is, they are no longer present as separate ions. However, the negative ions of the acid and the positive ions of the base are still present. Identify the importance of water to life by relating its functions as a transport and dissolving medium.

Water is very important to life. Some scientists beleived that life probaly evolved in water. All living cells are made up of 70-95% water and the water covers 75% of the earths surface. Water is present in all three physical states of mater - solid, liquid, and gas.

Water gets its properties from the fact that it is a polar covalent molecule, and therefore forms hydrogen bonds that connects one water molecule to as many as four other water molecules. This results in a property known as cohesion. Water also has adhesive properties to ionic substances as well as other charged substances such as clay. These properties result in many of water's important properties which include its remarkable ability to dissolve a vast variety of covalent and ionic substances, its high heat capacity, capillarity, surface tension, high heat of vaporization, and the fact that water floats when it freezes. All of these properties of water are related to life on Earth.

Water is considered the universal solvent of life. It acts as a solvent which transports molecules and ions from one area of a cell to another. It also acts as a reactant in chemical reactions that occur in the cell. Without water, molecules and ions would be unable to diffuse through the membrane that surrounds the cell, and gases such as carbon dioxide and oxygen would be unable to pass through plasma membranes.

Describe the Characteristics of Organic Compounds

Organic compounds contain carbon atoms that are linked by covalent bonds to form molecular chains, branches, rings, or sheets. Because there are 4 electrons in the outer energy level of the carbon atom, carbon can combine with other carbon atoms, as well as to the many other biologically important elements such as hydrogen, nitrogen, oxygen, sodium, magnesium, phosphorus, sulfur, chlorine, potassium, calcium, and iron. In addition to carbon, the vast majority of organic compounds also contain hydrogen. Oxides, carbonates and cyanides also contain carbon atoms but are not considered to be organic. Such compounds as carbon monoxide, carbon dioxide, calcium carbonate, and cyanide are considered to be inorganic, even though they contain carbon atoms.

Make an analysis of carbohydrates, lipids, proteins, and nucleic acids in terms of: the atoms that compose these compounds, the function these compounds serve in the body, the food sources of each compound, the sub-units that chemically combine to make these compounds, and examples of each group.

The first compound is carbohydrates. Carbohydrates are sugars and starches. Carbohydrates are found in breads, cereals, potatoes, pastas etc. Lipids is the next compound. Lipids are fats, oils and waxes. Lipids are found in mainly fried foods such as pizza, fries, etc. Protein is the next compound. Proteins are compounds that contain nitrogen as well as carbon, hydrogen and oxygen. Proteins are found in beans, fish, and meats. The final compound is nucleic acids. Nucleic acids contain phosphorous and nitrogen along with carbon, hydrogen and oxygen. Nucleic acids are mainly found in meats

Carbohydrates are compunds composed of carbon, hydrogen and oxygen. The name carbohydrates comes from the fact that they are formed from the union of carbon dioxide and water during the process of photosynthesis. The ratio of hydrogen to oxygen in a carbohydrate is always two to one - two hydrogen atoms for every oxygen atom. Carbohydrates are normally foun in the breads and cereals as well as fruits and vegetables.

The simplest carbohydrates are just simple sugars (called monosaccharides). These sugars are very important to the body, because they contain large amounts of energy. This energy can be released in the presence of oxygen by breaking down the sugars into carbon dioxide and water.

Lipids include subtances commonly known as fats, waxes and oils. Like carbohydrates, lipids are composed of carbon, hydrogen and oxygen, except there is less oxygen in lipids, than in carbohydrates. Lipids help make up parts of the cell structure (such as the cell membranes) and are used as a reserve for energy in an organism. Lipids provide twice as much energy per gram as the same amount of carbohydrate. Lipids are found in fatty meats such as bacon, as well as products such as whole milk, cream and butter, and from the oils of many plants such as olives, corn, and peanuts.

Proteins are compuounds containing nitrogen, carbon, hydrogen & oxygen. Some proteins also contain sulfur and phosphorus. The number of proteins in your body is virtually countless. Proteins are found everywhere, in cartilage, bones and muscles, hormones, antibodies, and even enzymes. Proteins are notmally found in meats, poltry, fish and some plants such as beans, soy beans, and legumes.

Nucleic Acids are organic compounds that contain Phosphorus, and Nitrogen, as well as Hydrogen, Oxygen and Carbon. Ther are only two known types of nucleic acids, and they are DNA (Deoxyribonucleic Acid) and RNA (Ribonucleic Acid). These two substances were first found in the Nucleus of a Cell. DNA (which is hereditary) and RNA direct and control the activities and structural devolpment of all the cells of an organism.

Specify the Importance of Nucleic Acids

Nucleic Acids are compounds that contain phosphorous and nitrogen in addition to carbon. There are two types of nucleic acids that are formed by the nucleolus located within the nucleus of cells. DNA (deoxyribonucleic acid) and RNA (ribonucleic acid) carry the genetic code that controls the manufacture or synthesis of proteins. These proteins serve as building blocks of cells, as hormones that regulate cellular function, as antibodies that fight infection, and as enzymes that regulate chemical reactions within the cell.

Nucleic acids are a group of organic substances found in the chromosomes of living cells and viruses. They play a central role in the storage and replication of hereditary information and in the expression of this information through protein synthesis. In most organisms, nucleic acids occur in combination with proteins; the combined substances are called nucleoproteins. Nucleic acid molecules are complex chains of varying length. The two chief types of nucleic acids are DNA (deoxyribonucleic acid), which carries the hereditary information from generation to generation, and RNA (ribonucleic acid), which delivers the instructions coded in this information to the cell's protein manufacturing sites (ribosomes).

Outline the function of minerals such as calcium, phosphorus, potassium, and sodium and describe why these substances are essential.

Minerals are chemical elements that organisms need for normal functioning. The most important minerals are iron, calcium, phosphorus, and iodine.

Calcium is the most abundant mineral in the human body, and for that matter, the body of any boned animal, as it is the structural element of bones. Weather our body is deficient in calcium depends less on whether we have a sufficient dietary intake, but rather on how well we absorb it. An important factor in calcium absorption is Vitamin D. The healthy body can produce this vitamin itself for as long as it gets some exposure to sunlight every day. Calcium absorption depends on hormones. Women beyond menopause are under increased risk of developing osteoporosis(bone loss), not because they would lacking in dietary calcium but because they lack the estrogen needed for proper absorption. Trace minerals such as boron also play a part in calcium absorption. The best sources of dietary calcium are milk and milk products, as well as soy and green leafy vegetables. Recommended daily amounts for dietary calcium are about 700 milligram for a healthy adult, and about a gram for adolescents.

Phosphates are essential to the energy-transfer reactions necessary to sustain life processes. Another name for this is adenosine triphospate or ATP which is highly significant to all cells in your body which is involved in every metabolic and photosynthetic reaction. In nature phosphorus occurs only in compounds called phosphates. Plants get phosphorus from soil, and use it in photosynthesis. People and animals take in phosphorus by eating plants and different types of foods.

Distinguish between carbohydrates (sugars and starches), lipids and proteins by using the appropriate tests to identify these nutrients. List each test and how it is conducted.

Tests for lipids: There are two tests that can be used to test for the presence of lipids:

1. Sudan IV : Sudna IV is a special biological stain that is soluble in nonpolar solvents such as lipids and turn from a pink to red colour. Polar compounds will not take up the pink or red colour. For example, if you take a test tube of water and vegetable oil (a lipid) and mix them together and then add a small amount of Sudan IV and shake the mixture, only the oil will turn red or pink not the water.

2. The second test is often called the translucence test for lipid. This is the test that uses unglazed brown paper (same type of brown paper you find in a brown paper bag like the ones from the Co-op used to rap bottles). If lipid is present, the lipid is absorbed into the paper and causes it to appear translucent which means that light is able to pass through and it looks like a greesy spot.

Tests for Protein: The Biuret Test is often used for protein. As you will learn soon, protein is made of a chain of amino acids that are bonded together by what are known as peptide bonds. The Biuret Test uses Biuret Reagent (a special solution) that is able to react with the peptide bonds of the amino acid chain that makes up the protein. This causes a colour change from blue indicating no protein, to pink (small amount of protein), to violet (larger amount of protein), to purple (largest amount of protein). An example to use is to test a known sample of protein such as egg white or gelatin. Adding the Biuret reagent to either of these food materials will cause a purple colour. Adding Biuret Reagent to water would result in blue colour.

Tests for carbohydrate: Benedicts Reagent is used for single sugars (called monosaccharides). Presence of monosaccharides causes the Benedict's solution to change to various colours when heated. The colour change depends on the concentration of the sugar present. Benedict's Reagent is made from Copper Sulfate and therefore contains Cu^{2+} ions. In the presence of simple sugars (known as reducing sugars) the copper is changed to Cu^{1+} ions. The reaction depends on the concentration of the sugar present. Here is a list of colour changes:

- Blue means no sugar present.
- Light green means 0.5% to 1.0% sugar present.
- Green to Yellow means 1.0% to 1.5% sugar present.
- Orange means 1.5% to 2.0% sugar present.
- Red to red brown means greater than 2.0% sugar present.

Iodine is the test for starch. Mixing iodine and starch causes the formation of a starch-iodine complex that is blue-black in colour. If you place iodine on a sample of food and it remains an amber or light brown colour there is no starch present, but if there is starch present, then the colour will change to a dark blue-black colour.

Explain what vitamins are and describe in general terms the vital body functions of vitamins A, B, C, D, E, and K.

Vitamins are organic molecules required in the diet in very small amounts. Vitamins are substances that are normally made by plants and then ingested by consumers. The vitamins are important because they serve as coenzymes which help the enzymes function more effectively. If the body is unable to supply those necessary enzymes in the proper quantities at the proper time, the vitamins and minerals become inert materials and pass unused through your body. Vitamins function directly as coenzymes, or are converted into coenzymes within the cell. Many important biological reactions need vitamins. Vitamin K and some of the B vitamins are essential for blood clotting. This is because Vitmain K is needed for the symthesis of prothrombin, one of the plasma proteins that takes part in the series of reactions leading to the formation of a blood clott.

When large doses of anitbiotics are given to overcome an infection, they can destroy the intestinal bacteria, and a Vitamin K deficiency may result. This is because Vitamin K is produced as a product of the metabolic activities of these bacteria.

Vitamin A

Vitamin A is a fat-soluble vitamin, which is structurally related to carotene. This is because carotene is converted into vitamin A in the liver. It is composed of compounds called retinoids and carotenoids, and is measured in Retinol Equivalents (RE's) which measure the vitamin A activity of food. There are many vital body functions in which vitamin A is used, some of these are maintaining the protective linings of the lungs, intestine, urinary tract, and other organs, as well as maintaining good vision, and promoting proper bone growth and tooth development.

Too little vitamin A leads to a conditon called night blindness. Night blind people have trouble seeing in dim light. Both black-and-white and colour vision make use of the light-sensitive pigment retinal, which is made from Vitamin A. Others include helping in the formation and maintenance of healthy skin and hair, healthy mucus membranes, fortifying the immune system, reproduction, and hormone synthesis and regulation.

Also, Vitamin A is used to treat minor skin disorders, it is important for the human metabolism, and is thought to help prevent the development of cancer. Sources of vitamin A include things such as some diary products, eggs, and liver. Some sources of carotene are orange fruits and vegetables, as well as dark, leafy greens. The absence of Vitamin A in ones diet can cause loss in weight, eye diseases (including night blindness), kidney stones, rashes, depression, frequent infections of the respiratory system, bladder, or digestive systems, plus many other problems.

Vitamin B1 (Thiamine)

Vitamin B1 is a water-soluble vitamin which is important for the production of energy and metabolism of sugar. It is also essential for supporting normal appetite, nervous system functions, muscles and heart, as well as aiding in the digestion of carbohydrates, and promoting growth and good muscle tone. Good sources of vitamin B1 are some dairy products, nuts, meats, whole grains, yeast eggs, and germ of cereals. The lack of vitamin B1 in ones diet may cause the loss of appetite, loss of weight, nausea, anxiety, irritability, insomnia, depression, weakness and fatigue. Other problems that may occur are heart and muscle problems, such as enlarged heart, abnormal heart rhythms, heart failure, muscle wasting, aches and pains in muscles, difficulty walking, loss of reflexes, and paralysis.

Vitamin B2 (Riboflavin)

Vitamin B2 is a water-soluble vitamin which is found to release energy from protein, fat, and carbohydrates, and is the precursor of flavoproteins, flavin-adenine dinucleotide and flavin mononucleotide. This vitamin is part of a coenzyme used in energy metabolism, it alleviates eye fatigue, helps maintain cell respiration, healthy eyes, hair, skin, and nails, aids in the formation of antibodies and red blood cells, and promotes general health. Good sources of vitamin B2 are some dairy products, liver, the white of an egg, meats, whole grains, cereals, almonds, and green vegetables. Some problems that are caused by the dietary lack of vitamin B2 are magenta tongue, rashes, cracks and sores in or around the mouth, itching, burning, bloodshot eyes, digestive disturbances, trembling, sluggishness, as well as dry, scaly skin.

Vitamin B3 (Niacin)

Vitamin B3 is a water-soluble vitamin, it's precursor is tryptophan, and it is involved in the oxidative release of energy from food. It is necessary for the metabolism of carbohydrates, is part of a coenzyme used in energy metabolism, helps improve circulation, supports the health of the nervous system, digestive system, and the skin, as well as protects the skin.

Other vital body functions of vitamin B3 are preventing pellagra, and reducing high blood pressure, and help lower cholesterol levels. Dietary sources of vitamin B3 include milk, eggs, meat, whole grains and cereals, nuts, and all protein-containing foods. Insufficient amounts of vitamin B3 in one's diet may result in pellagra, irritability, nervousness, loss of appetite, dizziness, mental confusion, flaky skin rash on sun exposed areas, and delirium. Others include depression, fatigue, headaches, muscular weakness, insomnia, and black, smooth tongue.

Vitamin B5 (Pantothenic Acid)

Vitamin B5 is a water-soluble vitamin which is essential for the metabolism of food. Some vital body functions of vitamin B5 include metabolism of proteins, carbohydrates, and fats, and the production of steroid hormones and other important chemicals, helps fight depression. As well as supporting the normal functioning of the gastrointestinal tract, and it is required for the production of cholesterol, bile, red blood cells, and antibodies. Good dietary sources of vitamin B5 are whole grains, beans, liver, eggs, fish, milk and milk products, yeast, lean beef, some potatoes. Others include broccoli, and other vegetables in the cabbage family. Deficiency of vitamin B5 causes problems such as abdominal distress, burning sensation in the heels, and sleep problems.

Vitamin B6 (Pyridoxine)

Vitamin B6 is a water-soluble vitamin, and is related compounds pyridoxamine, and pyridoxal. This vitamin is necessary for the synthesis and breakdown of amino acids, is used for the metabolism of protein, aids in the formation of red blood cells, helps convert tryptophan to niacin, and for maintaining the central nervous system. It also helps maintain a proper balance of sodium and phosphorous in the body, aids in the removal of excess fluid in premenstrual women, promotes healthy skin, and reduces muscle spasms, leg cramps, nausea, as well as hand numbness and stiffness.

Sources of vitamin B6 include bananas, fish, shellfish, meat, poultry, nuts, fruits, whole grains, vegetables, yeast, wheat germ, maize, and rice husks. The chance of deficiency is increased by smoking, and can cause problems such as anaemia, smooth tongue, insomnia, irritability, muscle twitching, convulsions, rashes, kidney stones, greasy dermatitis. In addition, other resulting problems are arm and leg cramps, loss of hair, muscular weakness, slow learning, mouth disorders, water retention, nervousness, and skin eruptions.

Vitamin B12 (Cyanocobalamin)

Vitamin B12 is a water-soluble vitamin, and is involved in the biosynthesis of methyl groups of choline, and methionine. Some vital body functions of vitamin B12 are being part of a coenzyme used in new cell synthesis, help maintain and protect the nerve cells and system, helps in the formation and regeneration of red blood cells, and helps prevent anaemia.

Others include being used for genetic materials, it is necessary for carbohydrate, fat and protein metabolism, promotes growth in children, increases energy, and is needed for calcium absorption. Significant sources of this vitamin are milk, cheese, eggs, liver and meat. A shortage of vitamin B12 in one's diet can lead to anaemia, poor appetite, growth failure in children, fatigue, depression, lack of balance, smooth tongue, and skin hypersensitivity. In addition, other problems are brain damage, paralysis, nervous system degeneration, as well as degeneration of the spinal cord.

Define Enzymes and Explain the Make-up and Importance of Enzymes

Enzymes (EN zymz) are protein substances that are necessary for most of the chemical reactions that occur in living cells. Enzymes speed up chemical reactions. Enzymes go into a chemical reaction only temporarily. They are not changed by this reaction so they may be used over and over again for the same chemical step with other molecules. In this way, enzymes are organic catalysts. The substance that an enzyme acts upon is called it's substrate.

Enzymes are almost invisible Protein molecules which are contained in the food we eat and are produced by our bodies. Enzymes are necessary for most of the chemical reations that occur in living cells. All enzymes in an organism are made by the cells of the organism. There are three different types of enzymes which are Metabolic, Digestive, and Food. Enzymes are important because they purify blood, strengthen the immune system, break down fats, lower cholesterol, enhance mental capacity, cleanse the colon, enhance sleep, and improve aging skin.

All living cells make enzymes, but enzymes are not alive. Enzyme molecules function by altering other molecules. Enzymes combine with the altered molecules to form a complex molecule structure in which chemical reactions take place. The enzyme, which remains unchanged, then seprates from the product of the reaction.

The human body has thousands of kinds of enzymes. Each kind does it's own job. Without these people could not breath, see, move, or digest food. Photosynthesis in plants also depends on enzymes.

There is a major difference between chemical reactions in nonliving things and the reactions in living things. Consider, for example, the burning of gasoline in an automobile engine. The gasoline vapor is admitted to the engine cylinder. It is ignited by a spark. The vapor burns in a fraction of a second. In fact, the burning is so rapid that it produces a small explosion, which helps to drive the engine.

How Enzymes work is that the ability to act as a catalysts depends on their shape. Somewhere on the surface of each enzyme there is a region called the active site. An Enzyme may also cause 2 molecules to join. In this case, there are two substrates. Each fits into the active site in such a way that they are brought into close contact. This enables bonds to form between the two substrate molecules. There is a theory to explain how enzyme and substrate fit together at an active site. It is called the lock-and-key model. Just as the notched surface of a key can open only one lock, the shape of the active site of an enzyme only fits certian substrates. Thus, each enzyme can catalyze a reaction only of those substrates.

The more modern view of enzyme function is known as the induced fit model which suggests that the enzyme changes shape just slightly to better fit the substrate.

Enzymes are proteins which act as biological catalysts. A catalyst enables a chemical reaction to take place at lower temperatures by lowering the activation energy required for the reaction to occur. This is important to living cells since the enzyme enables the metabolic reactions to occur at normal cellular temperatures. Many enzymes must be attached to certain non-protein molecules in order to function. Some molecules are called cofactors, such as copper, iron, and magnesium. Others are called coenzymes. Many coenzymes consist of vitamins, especially B vitamins. If a persons diet lacks the right amount of these vitamins, the enzymes cannot function properly, and serious problems may occur within the body.

Molecules of Life: Biomolecules

A biomolecule is any organic molecule that is produced by a living organism, including large polymeric molecules such as proteins, polysaccharides, and nucleic acids as well as small molecules such as primary metabolites, secondary metabolites, and natural products. A more general name for this class of molecules is a biogenic substance.

As organic molecules, biomolecules consist primarily of carbon and hydrogen, nitrogen, and oxygen, and, to a smaller extent, phosphorus and sulfur. Other elements sometimes are incorporated but are much less common.

Types of Biomolecules

A diverse range of biomolecules exist, including:

- Small molecules:

In the fields of pharmacology and biochemistry, a small molecule is a low molecular weight organic compound which is by definition not a polymer. The term small molecule, especially within the field of pharmacology, is usually restricted to a molecule that also binds with high affinity to a biopolymer such as protein, nucleic acid, or polysaccharide and in addition alters the activity or function of the biopolymer. The upper molecular weight limit for a small molecule is approximately 800 Daltons which allows for the possibility to rapidly diffuse across cell membranes so that they can reach intracellular sites of action. In addition, this molecular weight cutoff is a necessary but insufficient condition for oral bioavailability.

Small molecules can have a variety of biological functions, serving as cell signaling molecules, as tools in molecular biology, as drugs in medicine, as pesticides in farming, and in many other roles. These compounds can be natural (such as secondary metabolites) or artificial (such as antiviral drugs); they may have a beneficial effect against a disease (such as drugs) or may be detrimental (such as teratogens and carcinogens). Biopolymers such as nucleic acids, proteins, and polysaccharides (such as starch or cellulose) are not small molecules, although their constituent monomers — ribo- or deoxyribonucleotides, amino acids, and monosaccharides, respectively — are often considered to be. Very small oligomers are also usually considered small molecules, such as dinucleotides, peptides such as the antioxidant glutathione, and disaccharides such as sucrose.

Drugs

Most drugs are small molecules, although some drugs can be proteins, e.g. insulin. Many proteins are degraded if administered orally and most often cannot cross the cell membranes. Small molecules are more likely to be absorbed, although some of them are only absorbed after oral administration if given as prodrugs.

Many dietary supplements are small molecules (but not herb extracts, such as ginkgo).

Primary and Secondary Metabolites

For organisms to produce small molecules they need one or more specialized enzymes (to create and destroy), which as a result are not that varied in vertebrates (recent and small + slow population size), but very common in soil bacteria (such as streptomyces) and fungi, which in particular secrete antibiotics.

Plants also have several secondary metabolites, which play a role in cell signalling, pigmentation or in defence, several of which have also been used as drugs (medical and recreational):

- Alkaloids
- Glycosides
- Lipids

- Flavonoids
- Nonribosomal peptides, such as actinomycin-D
- Phenazines
- Phenols
- Polyketide
- Terpenes, including steroids
- Tetrapyrroles.

Research Tools

Enzymes and receptors are often activated or inhibited by endogenous protein, but can be also inhibited by endogenous or exogenous small molecule inhibitors or activators with can bind to the active site or on the allosteric site. An example is the teratogen and carcinogen phorbol 12-myristate 13-acetate which is a plant terpene which activates protein kinase C promoting cancer, making it a very useful investigative tool. There is also interest in creating small molecule artificial transcription factors to regulate gene expression, examples include wrenchnol (a wrench shaped molecule).

Binding of ligand can be characterised using a variety of analytical techniques such as surface plasmon resonance, microscale thermophoresis or dual polarisation interferometry to quantify the reaction affinities and kinetic properties and also any induced conformational change.

Nucleosides and Nucleotides

Nucleosides are molecules formed by attaching a nucleobase to a ribose ring. Examples of these include cytidine, uridine, adenosine, guanosine, thymidine and inosine.

Nucleosides can be phosphorylated by specific kinases in the cell, producing nucleotides. Both DNA and RNA are polymers, consisting of long, linear molecules. The repeating structural units, or monomers, of the nucleic acids are called nucleotides.

Each nucleotide is made of an acyclic nitrogenous base, a pentose and one to three phosphate groups. They contain carbon, nitrogen, oxygen, hydrogen and phosphorus. They serve as sources of chemical energy (adenosine triphosphate and guanosine triphosphate), participate in cellular signaling (cyclic guanosine monophosphate and cyclic adenosine monophosphate), and are incorporated into important cofactors of enzymatic reactions (coenzyme A, flavin adenine dinucleotide, flavin mononucleotide, and nicotinamide adenine dinucleotide phosphate).

Saccharides

Monosaccharides are the simplest form of carbohydrates with only one simple sugar. They essentially contain an aldehyde or ketone group in their structure. The presence of an aldehyde group in a monosaccharide is indicated by the prefix *aldo-*. Similarly, a ketone group is denoted by the prefix *keto-*.

Examples of monosaccharides are the hexoses glucose, fructose, and galactose and pentoses, ribose, and deoxyribose Consumed fructose and glucose have different rates of gastric emptying, are differentially absorbed and have different metabolic fates, providing multiple opportunities for 2 different saccharides to differentially affect food intake. Disaccharides are formed when two monosaccharides, or two single simple sugars, form a bond with removal of water. They can be hydrolyzed to yield their saccharin building blocks by boiling with dilute acid or reacting them with appropriate enzymes. Examples of disaccharides include sucrose, maltose, and lactose. Polysaccharides are polymerized monosaccharides, complex, carbohydrates. They have multiple simple sugars. Examples are starch, cellulose, and glycogen. They are generally large and often have a complex branched connectivity. Because of their size, polysaccharides are not water-soluble, but their many hydroxy groups become hydrated individually when exposed to water, and some polysaccharides form thick colloidal dispersions when heated in water.

Shorter polysaccharides, with 3 - 10 monomers, are called oligosaccharides. A fluorescent indicator-displacement molecular imprinting sensor was developed for discriminating saccharides. It successfully discriminated three brands of orange juice beverage. The change in fluorescence intensity of the sensing films resulting is directly related to the saccharide concentration.

Lignin

Lignin is a random polymer composed mainly of aromatic rings with short (up to three) aliphatic carbons chains connecting the rings. Lignin is the second most common biopolymer (after cellulose) and is one of the primary structural components of most plants.

It contains subunits derived from *p*-coumaryl alcohol, coniferyl alcohol, and sinapyl alcohol and is unusual among biomolecules in that it is racemic i.e. it is not optically active. The lack of optical activity is because the polymerization of lignin occurs via free radical coupling reactions in which there is no preference for either configuration at a chiral center.

Lipids

Lipids are chiefly fatty acid esters, and are the basic building blocks of biological membranes. Another biological role is energy storage (e.g., triglycerides). Most lipids consist of a polar or hydrophilic head (typically glycerol) and one to three nonpolar or hydrophobic fatty acid tails, and therefore they are amphiphilic. Fatty acids consist of unbranched chains of carbon atoms that are connected by single bonds alone (saturated fatty acids) or by both single and double bonds (unsaturated fatty acids). The chains are usually 14-24 carbon groups long, but it is always an even number.

For lipids present in biological membranes, the hydrophilic head is from one of three classes:

- Glycolipids, whose heads contain an oligosaccharide with 1-15 saccharide residues.
- Phospholipids, whose heads contain a positively charged group that is linked to the tail by a negatively charged phosphate group.
- Sterols, whose heads contain a planar steroid ring, for example, cholesterol.

Other lipids include prostaglandins and leukotrienes which are both 20-carbon fatty acyl units synthesized from arachidonic acid. They are also known as fatty acids

Amino Acids

Amino acids contain both amino and carboxylic acid functional groups. (In biochemistry, the term amino acid is used when referring to those amino acids in which the amino and carboxylate functionalities are attached to the same carbon, plus proline which is not actually an amino acid).

Amino acids are the building blocks of long polymer chains. With 2-10 amino acids such chains are called peptides, with 10-100 they are often called polypeptides, and longer chains are known as proteins. These protein structures have many structural and enzymatic roles in organisms.

There are twenty amino acids that are encoded by the standard genetic code, but there are more than 500 natural amino acids. When amino acids other than the set of twenty are observed in proteins, this is usually the result of modification after translation (protein synthesis). Only two amino acids other than the standard twenty are known to be incorporated into proteins during translation, in certain organisms:

- Selenocysteine is incorporated into some proteins at a UGA codon, which is normally a stop codon.
- Pyrrolysine is incorporated into some proteins at a UAG codon. For instance, in some methanogens in enzymes that are used to produce methane.

Besides those used in protein synthesis, other biologically important amino acids include carnitine (used in lipid transport within a cell), ornithine, GABA and taurine.

Protein Structure

The particular series of amino acids that form a protein is known as that protein's primary structure. This sequence is determined by the genetic makeup of the individual. Proteins have several, well-classified, elements of local structure formed by intermolecular attraction, this forms the secondary structure of protein.

They are broadly divided in two, alpha helix and beta sheet, also called beta pleated sheets. Alpha helices are formed of coiling of protein due to attraction between amine group of one amino acid with carboxylic acid group of other.

The coil contains about 3.6 amino acids per turn and the alkyl group of amino acid lie outside the plane of coil. Beta pleated sheets are formed by strong continuous hydrogen bond over the length of protein chain. Bonding may be parallel or antiparallel in nature.

Structurally, natural silk is formed of beta pleated sheets. Usually, a protein is formed by action of both these structures in variable ratios. Coiling may also be random. The overall 3D structure of a protein is termed its tertiary structure. It is formed as result of various forces like hydrogen bonding, disulfide bridges, hydrophobic interactions, hydrophilic interactions, van der Waals force etc. When two or more different polypeptide chains cluster to form a protein, quaternary structure of protein is formed. Quaternary structure is a unique attribute of polymeric and heteromeric proteins like hemoglobin, which consists of two alpha and two beta peptide chains.

Apoenzymes

An apoenzyme is the inactive storage and generally secretory form of a protein. This is required to protect the secretory cell from the activity of that protein. Apoenzymes becomes active enzyme on addition of a cofactor. Cofactors can be either inorganic (e.g., metal ions and iron-sulfur clusters) or organic compounds, (e.g., flavin and heme). Organic cofactors can be either prosthetic groups, which are tightly bound to an enzyme, or coenzymes, which are released from the enzyme's active site during the reaction.

Isoenzymes

Isoenzymes are enzymes with similar function but different structure. They are products of different genes. They are produced in different organs to perform the same function. LDH are examples of such enzymes. Their varied levels in blood are used to determine any deformity in the organ of secretion.

Vitamins

A vitamin is a compound that is generally not synthesized by a given organism but is nonetheless vital to its survival or health (for example coenzymes). These compounds must be absorbed, or eaten, but typically only in trace quantities. When originally proposed by Casimir Funk, a Polish biochemist, he believed them to all be basic and therefore named them vital amines. The "l" was later dropped to form the word vitamines.

4

Genetic Code

The genetic code is the set of rules by which information encoded in genetic material (DNA or mRNA sequences) is translated into proteins (amino acid sequences) by living cells. The code defines a mapping between tri-nucleotide sequences, called codons, and amino acids. With some exceptions, a triplet codon in a nucleic acid sequence specifies a single amino acid. Because the vast majority of genes are encoded with exactly the same code, this particular code is often referred to as the canonical or standard genetic code, or simply *the* genetic code, though in fact there are many variant codes. For example, protein synthesis in human mitochondria relies on a genetic code that differs from the standard genetic code.

Not all genetic information is stored using the genetic code. All organisms' DNA contains regulatory sequences, intergenic segments, and chromosomal structural areas that can contribute greatly to phenotype. Those elements operate under sets of rules that are distinct from the codon-to-amino acid paradigm underlying the genetic code.

Discovery

After the structure of DNA was discovered by James Watson and Francis Crick, who used the experimental evidence of Maurice Wilkins and Rosalind Franklin (among others), serious efforts to understand the nature of the encoding of proteins began.

The fact that codons consist of three DNA bases was first demonstrated in the Crick, Brenner et al. experiment. The first elucidation of a codon was done by Marshall Nirenberg and Heinrich J. Matthaei in 1961 at the National Institutes of Health. They used a cell-free system to translate a poly-uracil RNA sequence (i.e., UUUUU...) and discovered that the polypeptide that they had synthesized consisted of only the amino acid phenylalanine. They thereby deduced that the codon UUU specified the amino acid phenylalanine. This was followed by experiments in the laboratory of Severo Ochoa demonstrating that the poly-adenine RNA sequence (AAAAA...) coded for the polypeptide, poly-lysine. and the poly-cytosine RNA sequence (CCCCC...) coded for the polypeptide, poly-proline. Therefore the codon AAA specified the amino acid lysine, and the codon CCC specified the amino acid proline. Using different copolymers most of the remaining codons were then determined. Extending this work, Nirenberg and Philip Leder revealed the triplet nature of the genetic code and allowed the codons of the standard genetic code to be deciphered. In these experiments, various combinations of mRNA were passed through a filter which contained ribosomes, the components of cells that translate RNA into protein. Unique triplets promoted the binding of specific tRNAs to the ribosome. Leder and Nirenberg were able to determine the sequences of 54 out of 64 codons in their experiments.

Subsequent work by Har Gobind Khorana identified the rest of the genetic code. Shortly thereafter, Robert W. Holley determined the structure of transfer RNA (tRNA), the adapter molecule that facilitates the process of translating RNA into protein. This work was based upon earlier studies by Severo Ochoa, who received the Nobel prize in 1959 for his work on the enzymology of RNA synthesis. In 1968, Khorana, Holley and Nirenberg received the Nobel Prize in Physiology or Medicine for their work.

Transfer of Information Via the Genetic Code

The genome of an organism is inscribed in DNA, or in the case of some viruses, RNA. The portion of the genome that codes for a protein or an RNA is called a gene. Those genes that code for proteins are composed of tri-nucleotide units called codons, each coding for a single amino acid. Each nucleotide sub-unit consists of a phosphate, deoxyribose sugar and one of the 4 nitrogenous nucleobases. The purine bases adenine (A) and guanine (G) are larger and consist of two aromatic rings. The pyrimidine bases cytosine (C) and thymine (T) are smaller and consist of only one aromatic ring. In the double-helix configuration, two strands of DNA are joined to each other by hydrogen bonds in an arrangement known as base pairing. These bonds almost always form between an adenine base on one strand and a thymine on the other strand and between a cytosine base on one strand and a guanine base on the other. This means that the number of A and T residues will be the same in a given double helix, as will the number of G and C residues. In RNA, thymine (T) is replaced by uracil (U), and the deoxyribose is substituted by ribose.

Each protein-coding gene is transcribed into a template molecule of the related polymer RNA, known as messenger RNA or mRNA. This, in turn, is translated on the ribosome into an amino acid chain or polypeptide. The process of translation requires transfer RNAs specific for individual amino acids with the amino acids covalently attached to them, guanosine triphosphate as an energy source, and a number of translation factors. tRNAs have anticodons complementary to the codons in mRNA and can be "charged" covalently with amino acids at their 3' terminal CCA ends. Individual tRNAs are charged with specific amino acids by enzymes known as aminoacyl tRNA synthetases, which have high specificity for both their cognate amino acids and tRNAs. The high specificity of these enzymes is a major reason why the fidelity of protein translation is maintained.

There are $4^3 = 64$ different codon combinations possible with a triplet codon of three nucleotides; all 64 codons are assigned for either amino acids or stop signals during translation. If, for example, an RNA sequence, UUUAAACCC is considered and the reading frame starts with the first U (by convention, 5' to 3'), there are three codons, namely, UUU, AAA and CCC, each of which specifies one amino acid. This RNA sequence will be translated into an amino acid sequence, three amino acids long. A given amino acid may be encoded by between one and six different codon sequences. A comparison may be made with computer science, where the codon is similar to a word, which is the standard "chunk" for handling data (like one amino acid of a protein), and a nucleotide is similar to a bit, in that it is the smallest unit.

The standard genetic code is shown in the following tables. These are called forward and reverse codon tables, respectively. For example, the codon AAU represents the amino acid asparagine, and UGU and UGC represent cysteine (standard three-letter designations, Asn and Cys, respectively).

Salient Features

Sequence Reading Frame

A codon is defined by the initial nucleotide from which translation starts. For example, the string GGGAAACCC, if read from the first position, contains the codons GGG, AAA and CCC; and, if read from the second position, it contains the codons GGA and AAC; if read starting from the third position, GAA and ACC. Every sequence can thus be read in three reading frames, each of which will produce a different amino acid sequence (in the given example, Gly-Lys-Pro, Gly-Asn, or Glu-Thr, respectively). With double-stranded DNA there are six possible reading frames, three in the forward orientation on one strand and three reverse on the opposite strand. The actual frame in which a protein sequence is translated is defined by a start codon, usually the first AUG codon in the mRNA sequence.

Start/Stop Codons

Translation starts with a chain initiation codon (start codon). Unlike stop codons, the codon alone is not sufficient to begin the process. Nearby sequences (such as the Shine-Dalgarno sequence in *E. coli*) and initiation factors are also required to start translation. The most common start codon is AUG which is read as methionine or, in bacteria, as formylmethionine. Alternative start codons (depending on the organism), include "GUG" or "UUG"; these codons normally represent valine and leucine, respectively, but as a start codon, they are translated as methionine or formylmethionine.

The three stop codons have been given names: UAG is *amber*, UGA is *opal* (sometimes also called *umber*), and UAA is *ochre*. "Amber" was named by discoverers Richard Epstein and Charles Steinberg after their friend Harris Bernstein, whose last name means "amber" in German. The other two stop codons were named "ochre" and "opal" in order to keep the "color names" theme. Stop codons are also called "termination" or "nonsense" codons. They signal release of the nascent polypeptide from the ribosome because there is no cognate tRNA that has anticodons complementary to these stop signals, and so a release factor binds to the ribosome instead.

Effect of Mutations

During the process of DNA replication, errors occasionally occur in the polymerization of the second strand. These errors, called mutations, can have an impact on the phenotype of an organism, especially if they occur within the protein coding sequence of a gene. Error rates are usually very low—1 error in every 10–100 million bases—due to the "proofreading" ability of DNA polymerases.

Missense mutations and nonsense mutations are examples of point mutations, which can cause genetic diseases such as sickle-cell disease and thalassemia respectively. Clinically important missense mutations generally change the properties of the coded amino acid residue between being basic, acidic polar or non-polar, whereas nonsense mutations result in a stop codon.

Mutations that disrupt the reading frame sequence by indels (insertions or deletions) of a non-multiple of 3 nucleotide bases are known as frameshift mutations. These mutations usually result in a completely different translation from the original, and are also very likely to cause a stop codon to be read, which truncates the creation of the protein. These mutations may impair the function of the resulting protein, and are thus rare in *in vivo* protein-coding sequences. One reason inheritance of frameshift mutations is rare is that if the protein being translated is essential for growth under the selective pressures the organism faces, absence of a functional protein may cause death before the organism is viable. Frameshift mutations may result in severe genetic diseases such as Tay-Sachs disease.

Although most mutations that change protein sequences are harmful or neutral, some mutations have a positive effect on an organism. These mutations may enable the mutant organism to withstand particular environmental stresses better than wild-type organisms, or reproduce more quickly. In these cases a mutation will tend to become more common in a population through natural selection. Viruses that use RNA as their genetic material have rapid mutation rates, which can be an advantage since these viruses will evolve constantly and rapidly, and thus evade the defensive responses of e.g. the human immune system. In large populations of asexually reproducing organisms, for example, *E. coli*, multiple beneficial mutations may co-occur. This phenomenon is called clonal interference and causes competition among the mutations.

Degeneracy

Degeneracy is the redundancy of the genetic code. The genetic code has redundancy but no ambiguity. For example, although codons GAA and GAG both specify glutamic acid (redundancy), neither of them specifies any other amino acid (no ambiguity). The codons encoding one amino acid may differ in any of their three positions. For example the amino acid glutamic acid is specified by GAA and GAG codons (difference in the third position), the amino acid leucine is specified by UUA, UUG, CUU, CUC, CUA, CUG codons (difference in the first or third position), while the amino acid serine is specified by UCA, UCG, UCC, UCU, AGU, AGC (difference in the first, second or third position).

A position of a codon is said to be a fourfold degenerate site if any nucleotide at this position specifies the same amino acid. For example, the third position of the glycine codons (GGA, GGG, GGC, GGU) is a fourfold degenerate site, because all nucleotide substitutions at this site are synonymous; i.e., they do not change the amino acid. Only the third positions of some codons may be fourfold degenerate. A position of a codon is said to be a twofold degenerate site if only two of four possible nucleotides at this position specify the same amino acid. For example, the third position of the glutamic acid codons (GAA, GAG) is a twofold degenerate site. In twofold degenerate sites, the equivalent nucleotides are always either two purines (A/G) or two pyrimidines (C/U), so only transversional substitutions (purine to pyrimidine or pyrimidine to purine) in twofold degenerate sites are nonsynonymous. A position of a codon is said to be a non-degenerate site if any mutation at this position results in amino acid substitution. There is only one threefold degenerate site where changing to three of the four nucleotides may have no effect on the amino acid (depending on what it is changed to), while changing to the fourth possible nucleotide always results in an amino acid substitution. This is the third position of an isoleucine codon: AUU, AUC, or AUA all encode isoleucine, but AUG encodes methionine. In computation this position is often treated as a twofold degenerate site.

There are three amino acids encoded by six different codons: serine, leucine, and arginine. Only two amino acids are specified by a single codon. One of these is the amino-acid methionine, specified by the codon AUG, which also specifies the start of translation; the other is tryptophan, specified by the codon UGG. The degeneracy of the genetic code is what accounts for the existence of synonymous mutations.

Degeneracy results because there are more codons than encodable amino acids. For example, if there were two bases per codon, then only 16 amino acids could be coded for ($4^2=16$). Because at least 21 codes are required (20 amino acids plus stop), and the next largest number of bases is three, then 4^3 gives 64 possible codons, meaning that some degeneracy must exist.

These properties of the genetic code make it more fault-tolerant for point mutations. For example, in theory, fourfold degenerate codons can tolerate any point mutation at the third position, although codon usage bias restricts this in practice in many organisms; twofold degenerate codons can tolerate one out of the three possible point mutations at the third position.

Since transition mutations (purine to purine or pyrimidine to pyrimidine mutations) are more likely than transversion (purine to pyrimidine or vice-versa) mutations, the equivalence of purines or that of pyrimidines at twofold degenerate sites adds a further fault-tolerance. A practical consequence of redundancy is that some errors in the genetic code only cause a silent mutation or an error that would not affect the protein because the hydrophilicity or hydrophobicity is maintained by equivalent substitution of amino acids; for example, a codon of NUN (where N = any nucleotide) tends to code for hydrophobic amino acids. NCN yields amino acid residues that are small in size and moderate in hydropathy; NAN encodes average size hydrophilic residues. These tendencies may result from the shared ancestry of the aminoacyl tRNA synthetases related to these codons.

Despite the redundancy of the genetic code, single point mutations can still cause dysfunctional proteins. For example, a mutated hemoglobin gene causes sickle-cell disease. In the mutant hemoglobin a hydrophilic glutamate (Glu) is substituted by the hydrophobic valine (Val), that is, GAA or GAG becomes GUA or GUG.

The substitution of glutamate by valine reduces the solubility of β-globin which causes hemoglobin to form linear polymers linked by the hydrophobic interaction between the valine groups, causing sickle-cell deformation of erythrocytes. Sickle-cell disease is generally not caused by a *de novo* mutation. Rather it is selected for in geographic regions where malaria is common (in a way similar to thalassemia), as heterozygous people have some resistance to the malarial *Plasmodium* parasite (heterozygote advantage).

These variable codes for amino acids are allowed because of modified bases in the first base of the anticodon of the tRNA, and the base-pair formed is called a wobble base pair. The modified bases include inosine and the Non-Watson-Crick U-G basepair.

Variations to the Standard Genetic Code

While slight variations on the standard code had been predicted earlier, none were discovered until 1979, when researchers studying human mitochondrial genes discovered they used an alternative code.

Many slight variants have been discovered since then, including various alternative mitochondrial codes, and small variants such as translation of the codon UGA as tryptophan in the species *Mycoplasma* and translation of CUG as a serine rather than a leucine in the genus *Candida*. In bacteria and archaea, GUG and UUG are common start codons, but in rare cases, certain proteins may use alternative start codons not normally used by that species.

In certain proteins, non-standard amino acids are substituted for standard stop codons, depending on associated signal sequences in the messenger RNA. For example, UGA can code for selenocysteine, and UAG can code for pyrrolysine. Selenocysteine is now viewed as the 21st amino acid, and pyrrolysine is viewed as the 22nd.

Despite these differences, all known codes are very similar to each other, and the coding mechanism is the same for all organisms: three-base codons, tRNA, ribosomes, reading the code in the same direction and translating the code three letters at a time into sequences of amino acids.

Expanded Genetic Code

Since 2001, 40 non-natural amino acids have been added into protein by creating a unique codon (recoding) and a corresponding transfer-RNA:aminoacyl – tRNA-synthetase pair to encode it with diverse physicochemical and biological properties in order to be used as a tool to exploring protein structure and function or to create novel or enhanced proteins.

An expanded genetic code refers to an artificially modified genetic code in which one or more specific codons have been allocated to encode an amino acid which is not among the twenty/twenty-two found in nature.

Background

The translation is catalysed by ribosomes. Transfer RNAs (tRNA) are used as keys to decode the RNA into its equivalent polypeptide. The tRNA recognises a specific three nucleotide codon thanks to a complementary sequence called the anticodon on one of its loops.

Each three nucleotide codon is translated into one amino acid. There is at least one tRNA for any codon. If there are more than one, they code for the same amino acid. Many tRNAs are compatible with several codons. The encoding of a codon to its amino acid is a result of the aminoacyl tRNA synthetase which adds the aminoacyl group to its allocated tRNA.

The aminoacyl tRNA synthetase often does not recognise the anticodon, but another part of the tRNA, meaning that if the anticodon were to be mutated the encoding of that amino acid would change to a new codon.

For successful translation of a novel amino acid, the codon to which the amino acid is reassigned must be free or unfavoured and the novel tRNA and synthetase set (called the orthogonal set when including the codon) must not crosstalk with the endogenous tRNA and synthetase sets, while still being functionally compatible with the ribosome and other components of the translation apparatus. The tRNA synthetase pair is taken from a distant organism, generally from a different domain, and the active site of the synthetase is modified to accept the non-natural amino acid.

The possibility of reassigning codons was realized by Normanly et al. in 1990 when a viable mutant strain of E. coli read through the amber (stop) codon. As a result the amber codon became the choice codon to be assigned a novel amino acid. Later, in the Schultz lab the tRNATyr/tyrosyl-tRNA synthetase (TyrRS) from Methanococcus jannaschii was used to introduce a tyrosine instead of STOP, the default value of the amber codon.

As mentioned, this was possible because of the differences between the endogenous bacterial synthases and the orthologous archeal synthase which do not recognise each other.

Directed Evolution

This orthologous set can then be mutated and screened through directed evolution to accept a different, even novel, amino acid. Mutations to the plasmid containing the pair can be introduced by error-prone PCR or through degenerate primers for the synthetase's active site. Selection involves multiple rounds of a two-step process, where the plasmid is transferred into cells expressing chloramphenicol acetyl transferase with a premature amber codon. In the presence of toxic chloramphenicol and the non-natural amino acid, the surviving cells will have overridden the amber codon using the orthogonal tRNA aminoacylated with either the standard amino acids or the non-natural one. To remove the former, the plasmid is inserted into cells with a barnase gene (toxic) with a premature amber codon but without the non-natural amino acid, removing all the orthogonal synthases which do not specifically recognize the non-natural amino acid. In addition to the recoding of the tRNA to a different codon, they can be mutated to recognize a four base codon, allowing additional free coding options. The non natural amino acid, as a result, introduces diverse physicochemical and biological properties in order to be used as a tool to explore protein structure and function or to create novel or enhanced protein for practical purposes.

Diversity

The orthogonal pairs of synthase and tRNA which work for one organism may not work for another as the synthase may mis-aminoacylate endogenous tRNAs or the tRNA be mis-aminoacylated itself by an endogenous synthase. As a result the sets created to date differ between organisms.

Orthogonal Sets in E. coli

- tRNA^{Tyr}-TyrRS pair from the archaeon *Methanococcus jannaschii*
- tRNA^{Lys}–LysRS pair from the archaeon *Pyrococcus horikoshii*
- tRNA^{Glu}–GluRS pair from *Methanosarcina mazei*
- leucyl-tRNA synthetase from *Methanobacterium thermoautotrophicum* and a mutant leucyl tRNA derived from *Halobacterium* sp.

Orthogonal Sets in Yeast

- tRNA^{Tyr}-TyrRS pair from *Escherichia coli*
- tRNA^{Leu}–LeuRS pair from *Escherichia coli*
- tRNA^{iMet} from human and GlnRS from *Escherichia coli*.

Orthogonal Sets in Mammalian Cells

- tRNA^{Tyr}-TyrRS pair from *Bacillus stearothermophilus*
- modified tRNA^{Trp}-TrpRS pair from *Bacillus subtilis* trp
- tRNA^{Leu}–LeuRS pair from *Escherichia coli*.

Protein Studies

With an expanded genetic code, the unnatural amino acid can be genetically directed to any chosen site in the protein of interest. The high efficiency and fidelity allows a better control of the placement of the modification compared to modifying the protein post-translationally, which generally will target all amino acids of the same type, such as the thiol group of cysteine and the -amino group of lysine. Also, an expanded genetic code allows modifications to be carried out *in vivo*. The ability to site-specifically direct lab-synthesized chemical moieties into proteins allows many types of studies which would otherwise be extremely difficult.

- Probing Protein Structure and Function: by using amino acids with slightly different size such as o-Methyltyrosine or dansylalanine instead of tyrosine, and by inserting genetically-coded reporter moieties (color-changing and/or spin-active) into selected protein sites, chemical information about the protein's structure and function can be measured.

- Identifying and Regulating Protein Activity: by using photocaged aminoacids, protein function can be "switched" on or off by illuminating the organism.

- Changing the mode of action of a protein: one can start with the gene for a protein which binds a certain sequence of DNA, and, by inserting a chemically active amino acid into the binding site, convert it to a protein which cuts the DNA, rather than binding it.

- Improving immunogenicity and overcoming self-tolerance: by replacing strategically-chosen tyrosines with *p*-nitro phenylalanine, a tolerated self-protein can be made immunogenic.

An example of the possible application for this method is the biomedical where "chemical warheads" can be added to protein which target specific cellular components.

Origin

Despite the minor variations that exist, the genetic code used by all known forms of life is nearly universal. However, there are a huge number of possible genetic codes. If amino acids are randomly associated with triplet codons, there will be 1.5×10^{84} possible genetic codes. Phylogenetic analysis of transfer RNA suggests that tRNA molecules evolved before the present set of aminoacyl-tRNA synthetases.

Theoretically, the genetic code could be completely random (a "frozen accident"), completely non-random (optimal) or a combination of random and nonrandom. There are enough data to refute the first possibility. For a start, a quick view on the table of the genetic code shows a clustering of amino acid assignments. Furthermore, amino acids that share the same biosynthetic pathway tend to have the same first base in their codons, and amino acids with similar physical properties tend to have similar codons.

There are four themes running through the many theories about the evolution of the genetic code (and hence the origin of these patterns):

- Chemical principles govern specific RNA interaction with amino acids. Experiments with aptamers showed that some amino acids have a selective chemical affinity for the base triplets that code for them. Recent experiments show that of the 8 amino acids tested, 6 show some RNA triplet-amino acid association. This has been called the stereochemical code. The stereochemical code could have created an ancient core of assignments. The current complex translation mechanism involving tRNA and associated enzymes may be a later development, and maybe protein sequences were directly templated on base sequences.

- Biosynthetic expansion. The standard modern genetic code grew from a simpler earlier code through a process of "biosynthetic expansion". Here the idea is that primordial life "discovered" new amino acids (for example, as by-products of metabolism) and later incorporated some of these into the machinery of genetic coding. Although much circumstantial evidence has been found to suggest that fewer different amino acids were used in the past than today, precise and detailed hypotheses about which amino acids entered the code in what order have proved far more controversial.

- Natural selection has led to codon assignments of the genetic code that minimize the effects of mutations. A recent hypothesis suggests that the triplet code was derived from codes that used longer than triplet codons. Longer than triplet decoding has higher degree of codon redundancy and is more error resistant than the triplet decoding. This feature could allow accurate decoding in the absence of highly complex translational machinery such as the ribosome.

- Information channels: Information-theoretic approaches see the genetic code as an error-prone information channel. The inherent noise (that is, errors) in the channel poses the organism with a fundamental question: how to construct a genetic code that can withstand the impact of noise while accurately and efficiently translating information? These "rate-distortion" models suggest that the genetic code originated as a result of the interplay of the three conflicting evolutionary forces: the needs for diverse amino-acids, for error-tolerance and for minimal cost of resources. The code emerges at a coding transition when the mapping of codons to amino-acids becomes nonrandom. The emergence of the code is governed by the topology defined by the probable errors and is related to the map coloring problem.

Gene: Its Expression

Gene expression is the process by which information from a gene is used in the synthesis of a functional gene product. These products are often proteins, but in non-protein coding genes such as ribosomal RNA (rRNA) genes or transfer RNA (tRNA) genes, the product is a functional RNA. The process of gene expression is used by all known life - eukaryotes (including multicellular organisms), prokaryotes (bacteria and archaea) and viruses - to generate the macromolecular machinery for life. Several steps in the gene expression process may be modulated, including the transcription, RNA splicing, translation, and post-translational modification of a protein. Gene regulation gives the cell control over structure and function, and is the basis for cellular differentiation, morphogenesis and the versatility and adaptability of any organism. Gene regulation may also serve as a substrate for evolutionary change, since control of the timing, location, and amount of gene expression can have a profound effect on the functions (actions) of the gene in a cell or in a multicellular organism.

In genetics, gene expression is the most fundamental level at which the genotype gives rise to the phenotype. The genetic code stored in DNA is "interpreted" by gene expression, and the properties of the expression give rise to the organism's phenotype.

Mechanism

Transcription

The gene itself is typically a long stretch of DNA which carries genetic information encoded by genetic code. Every molecule of DNA consists of two strands, each of them having 5' and 3' ends oriented in anti-parallel direction. The coding strand contains the genetic information while template strand (non-coding strand) serves as a blueprint for the production of RNA. The production of RNA copies of the DNA is called transcription, and is performed by RNA polymerase, which adds one RNA nucleotide at a time to a growing

RNA strand. This RNA is complementary to the template 3' '! 5' DNA strand, which is itself complementary to the coding 5' '! 3' DNA strand. Therefore, the resulting 5' '! 3' RNA strand is identical to the coding DNA strand with the exception that thymines (T) are replaced with uracils (U) in the RNA. A coding DNA strand reading "ATG" is transcribed as "AUG" in RNA.

Transcription in prokaryotes is carried out by a single type of RNA polymerase, which needs DNA sequence called Pribnow box and sigma factor (σ factor) to start transcription. In eukaryotes, the transcription is done by three types of RNA polymerases, each of them needs special DNA sequence called promoter and a set of DNA-binding proteins - transcription factors to initiate the process. RNA polymerase I is responsible for transcription of rRNA genes, while RNA polymerase II transcribes all protein-coding genes but also some non-coding RNAs (e.g. snRNAs, snoRNAs or long non-coding RNAs) as well. It contains special part called C-terminal domain (CTD) that is rich of serines, which after being phosphorylated accumulate factors necessary for RNA modification and maturation. RNA polymerase III transcribes 5S rRNA and tRNA genes but also some small non-coding RNA genes (e.g. 7SK). Transcription ends on a special sequence called terminator.

RNA Processing

While transcription of prokaryotic protein-coding genes creates messenger RNA (mRNA) which is ready for translation, transcription of eukaryotic genes leaves a primary transcript of RNA (pre-mRNA), which first has to undergo series of modification to become a mature mRNA.

These include 5' *capping*, which is set of enzymatic reactions that add 7-methylguanosine (m^7G) to the 5' end of pre-mRNA and thus protect the RNA from degradation by exonucleases. The m^7G cap is then bound by cap binding complex heterodimer (CBC20/CBC80) which aids in mRNA export to cytoplasm and also protect the RNA from decapping.

Another modification is 3' *clevage and polyadenylation*. They occur if polyadenylation signal sequence (5'- AAUAAA-3') is present in pre-mRNA,which is usually between protein-coding sequence and terminator. The pre-mRNA is first cleaved and then a series of ~200 adenines (A) are added to form poly(A) tail which protects the RNA from degradation. Poly(A) tail is bound by multiple poly(A)-binding proteins (PABP) necessary for mRNA export and translation re-iniciation.

Very important modification of eukaryotic pre-mRNA is *RNA splicing.* Majority of eukaryotic pre-mRNAs consist of alternating segments called exons and introns. During the process of splicing, RNA-protein catalytical complex known as spliceosome, catalyze two transesterification reactions, which remove intron and release it in form of lariat structure and then splice neighbouring exons together. In certain cases, some introns or exons can be either removed or retained in mature mRNA. This so-called alternative splicing creates series of different transcripts originating from a single gene. Because these transcripts can be potentially translated into different proteins, splicing extends the complexity of eukaryotic gene expression.

Extensive RNA processing may be an evolutionary advantage made possible by the nucleus of eukaryotes. In prokaryotes transcription and translation happen together whilst in eukaryotes the nuclear membrane separates the two processes giving time for RNA processing to occur.

Non-coding RNA Maturation

In most organisms non-coding genes (ncRNA) are transcribed as precursors which undergo further processing. In the case of ribosomal RNAs (rRNA), they are often transcribed as a pre-rRNA which contains one or more rRNAs, the pre-rRNA is cleaved and modified (22 -O-methylation and pseudouridine formation) at a specific sites by approximately 150 different small nucleolus-restricted RNA species, called snoRNAs. SnoRNAs associate with proteins, forming snoRNPs. While snoRNA part basepair with the target RNA and thus position the modification to precise site, the protein part performs the catalytical reaction. In eukaryotes, in particular a snoRNP, called RNase MRP cleaves the 45S pre-rRNA into the 28S, 5.8S, and 18S rRNAs. The rRNA and RNA processing factors form large aggregates called the nucleolus.

In the case of transfer RNA (tRNA), for example, the 5' sequence is removed by RNase P, whereas the 3' end is removed by the tRNase Z enzyme and the non-templated 3' CCA tail is added by a nucleotidyl transferase. In the case of micro RNA (miRNA), miRNAs are first transcribed as primary transcripts or pri-miRNA with a cap and poly-A tail and processed to short, 70-nucleotide stem-loop structures known as pre-miRNA in the cell nucleus by the enzymes Drosha and

Pasha. After being exported, it is then processed to mature miRNAs in the cytoplasm by interaction with the endonuclease Dicer, which also initiates the formation of the RNA-induced silencing complex (RISC), composed of the Argonaute protein. Even snRNAs and snoRNAs themselves undergo series of modification before they become part of functional RNP complex. This is done either in the nucleoplasm or in the specialized compartments called Cajal bodies. Their bases are methylated or pseudouridinilated by a group of small Cajal body-specific RNAs (scaRNAs) which are structurally similar to snoRNAs.

RNA Export

In eukaryotes most mature RNA must be exported to the cytoplasm from the nucleus. While some RNAs function in the nucleus, many RNAs are transported through the nuclear pores and into the cytosol. Notably this includes all RNA types involved in protein synthesis. In some cases RNAs are additionally transported to a specific part of the cytoplasm, such as a synapse; they are then towed by motor proteins that bind through linker proteins to specific sequences (called "zipcodes") on the RNA.

Translation

For some RNA (non-coding RNA) the mature RNA is the final gene product. In the case of messenger RNA (mRNA) the RNA is an information carrier coding for the synthesis of one or more proteins. mRNA carrying a single protein sequence (common in eukaryotes) is monocistronic whilst mRNA carrying multiple protein sequences (common in prokaryotes) is known as polycistronic.

Every mRNA consists of three parts - 5' untranslated region (5'UTR), protein-coding region or open reading frame (ORF) and 3' untranslated region (3'UTR). Coding region carries information for protein synthesis encoded by genetic code into form of triplets. Each triplet of nucleotides of the coding region is called codon and corresponds to a binding site complementary to an anticodon triplet in transfer RNA. Transfer RNAs with the same anticodon sequence always carry identical type of amino acid. Amino acids are then chained together by the ribosome according to order of triplets in the coding region. The ribosome helps transfer RNA to bind to messenger RNA and takes the amino acid from each transfer RNA and makes a structure-less protein out of it.

In prokaryotes translation generally occurs at the point of transcription (co-transcriptionally), often using a messenger RNA which is still in the process of being created. In eukaryotes translation can occur in a variety of regions of the cell depending on where the protein being written is supposed to be. Major locations are the cytoplasm for soluble cytoplasmic proteins and the membrane of endoplasmic reticulum for proteins which are for export from the cell or insertion into a cell membrane. Proteins which are supposed to be expressed at the endoplasmic reticulum are recognised part-way through the translation process. This is governed by the signal recognition particle - a protein which binds to the ribosome and directs it to the endoplasmic reticulum when it finds a signal sequence on the growing (nascent) amino acid chain.

Folding

The polypeptide folds into its characteristic and functional three-dimensional structure from random coil. Each protein exists as an unfolded polypeptide or random coil when translated from a sequence of mRNA to a linear chain of amino acids. This polypeptide lacks any developed three-dimensional structure. Amino acids interact with each other to produce a well-defined three-dimensional structure, the folded protein, known as the native state. The resulting three-dimensional structure is determined by the amino acid sequence (Anfinsen's dogma).

The correct three-dimensional structure is essential to function, although some parts of functional proteins may remain unfolded Failure to fold into the intended shape usually produces inactive proteins with different properties including toxic prions. Several neurodegenerative and other diseases are believed to result from the accumulation of *misfolded* (incorrectly folded) proteins. Many allergies are caused by the folding of the proteins, for the immune system does not produce antibodies for certain protein structures. Enzymes called chaperones assist the newly formed protein to attain (fold into) the 3-dimensional structure it needs to function. Similarly, RNA chaperones help RNAs attain their functional shapes. Assisting protein folding is one of the main roles of the endoplasmic reticulum in eukaryotes.

Protein Transport

Many proteins are destined for other parts of the cell than the cytosol and a wide range of signalling sequences are used to direct proteins to where they are supposed to be. In prokaryotes this is normally a simple process due to limited compartmentalisation of the cell. However in eukaryotes there is a great variety of different targeting processes to ensure the protein arrives at the correct organelle.

Not all proteins remain within the cell and many are exported, for example digestive enzymes, hormones and extracellular matrix proteins. In eukaryotes the export pathway is well developed and the main mechanism for the export of these proteins is translocation to the endoplasmic reticulum, followed by transport via the Golgi apparatus.

Regulation of Gene Expression

Regulation of gene expression refers to the control of the amount and timing of appearance of the functional product of a gene. Control of expression is vital to allow a cell to produce the gene products it needs when it needs them; in turn this gives cells the flexibility to adapt to a variable environment, external signals, damage to the cell, etc. Some simple examples of where gene expression is important are:

- Control of Insulin expression so it gives a signal for blood glucose regulation
- X chromosome inactivation in female mammals to prevent an "overdose" of the genes it contains.
- Cyclin expression levels control progression through the eukaryotic cell cycle.

More generally gene regulation gives the cell control over all structure and function, and is the basis for cellular differentiation, morphogenesis and the versatility and adaptability of any organism.

Any step of gene expression may be modulated, from the DNA-RNA transcription step to post-translational modification of a protein. The stability of the final gene product, whether it is RNA or protein, also contributes to the expression level of the gene - an unstable product results in a low expression level. In general gene expression is regulated through changes in the number and type of interactions between molecules that collectively influence transcription of DNA and translation of RNA.

Numerous terms are used to describe types of genes depending on how they are regulated, these include:

- A *constitutive gene* is a gene that is transcribed continually compared to a facultative gene which is only transcribed when needed.

- A *housekeeping gene* is typically a constitutive gene that is transcribed at a relatively constant level. The housekeeping gene's products are typically needed for maintenance of the cell. It is generally assumed that their expression is unaffected by experimental conditions. Examples include actin, GAPDH and ubiquitin.

- A *facultative gene* is a gene which is only transcribed when needed compared to a constitutive gene.

- An *inducible gene* is a gene whose expression is either responsive to environmental change or dependent on the position in the cell cycle.

Transcriptional Regulation

Regulation of transcription can be broken down into three main routes of influence; genetic (direct interaction of a control factor with the gene), modulation (interaction of a control factor with the transcription machinery) and epigenetic (non-sequence changes in DNA structure which influence transcription).

Direct interaction with DNA is the simplest and the most direct method by which a protein can change transcription levels. Genes often have several protein binding sites around the coding region with the specific function of regulating transcription. There are many classes of regulatory DNA binding sites known as enhancers, insulators, repressors and silencers. The mechanisms for regulating transcription are very varied, from blocking key binding sites on the DNA for RNA polymerase to acting as an activator and promoting transcription by assisting RNA polymerase binding.

The activity of transcription factors is further modulated by intracellular signals causing protein post-translational modification including phosphorylated, acetylated, or glycosylated. These changes influence a transcription factor's ability to bind, directly or indirectly, to promoter DNA, to recruit RNA polymerase, or to favor elongation of a newly synthetized RNA molecule.

The nuclear membrane in eukaryotes allows further regulation of transcription factors by the duration of their presence in the nucleus which is regulated by reversible changes in their structure and by binding of other proteins. Environmental stimuli or endocrine signals may cause modification of regulatory proteins eliciting cascades of intracellular signals, which result in regulation of gene expression.

More recently it has become apparent that there is a huge influence of non-DNA-sequence specific effects on translation. These effects are referred to as epigenetic and involve the higher order structure of DNA, non-sequence specific DNA binding proteins and chemical modification of DNA. In general epigenetic effects alter the accessibility of DNA to proteins and so modulate transcription.

DNA methylation is a widespread mechanism for epigenetic influence on gene expression and is seen in bacteria and eukaryotes and has roles in heritable transcription silencing and transcription regulation. In eukaryotes the structure of chromatin, controlled by the histone code, regulates access to DNA with significant impacts on the expression of genes in euchromatin and heterochromatin areas.

Post-transcriptional Regulation

In eukaryotes, where export of RNA is required before translation is possible, nuclear export is thought to provide additional control over gene expression. All transport in and out of the nucleus is via the nuclear pore and transport is controlled by a wide range of importin and exportin proteins.

Expression of a gene coding for a protein is only possible if the messenger RNA carrying the code survives long enough to be translated. In a typical cell an RNA molecule is only stable if specifically protected from degradation. RNA degradation has particular importance in regulation of expression in eukaryotic cells where mRNA has to travel significant distances before being translated. In eukaryotes RNA is stabilised by certain post-transcriptional modifications, particularly the 5' cap and poly-adenylated tail.

Intentional degradation of mRNA is used not just as a defence mechanism from foreign RNA (normally from viruses) but also as a route of mRNA *destabilisation*. If an mRNA molecule has a complementary sequence to a small interfering RNA then it is targeted for destruction via the RNA interference pathway.

Translational Regulation

Direct regulation of translation is less prevalent than control of transcription or mRNA stability but is occasionally used. Inhibition of protein translation is a major target for toxins and antibiotics in order to kill a cell by overriding its normal gene expression control. Protein synthesis inhibitors include the antibiotic neomycin and the toxin ricin.

Protein Degradation

Once protein synthesis is complete the level of expression of that protein can be reduced by protein degradation. There are major protein degradation pathways in all prokaryotes and eukaryotes of which the proteasome is a common component. An unneeded or damaged protein is often labelled for degradation by addition of ubiquitin.

Measurement

Measuring gene expression is an important part of many life sciences - the ability to quantify the level at which a particular gene is expressed within a cell, tissue or organism can give a huge amount of information. For example measuring gene expression can:

- Identify viral infection of a cell (viral protein expression)
- Determine an individual's susceptibility to cancer (oncogene expression)
- Find if a bacterium is resistant to penicillin (beta-lactamase expression).

Similarly the analysis of the location of expression protein is a powerful tool and this can be done on an organism or cellular scale. Investigation of localisation is particularly important for study of development in multicellular organisms and as an indicator of protein function in single cells. Ideally measurement of expression is done by detecting the final gene product (for many genes this is the protein) however it is often easier to detect one of the precursors, typically mRNA, and infer gene expression level.

mRNA Quantification

Levels of mRNA can be quantitatively measured by Northern blotting which gives size and sequence information about the mRNA molecules. A sample of RNA is separated on an agarose gel and hybridized to a radio-labeled RNA probe that is complementary to the target sequence. The radio-labeled RNA is then detected by an autoradiograph. The main problems with Northern blotting stem from the use of radioactive reagents (which make the procedure time consuming and potentially dangerous) and lower quality quantification than more modern methods (due to the fact that quantification is done by measuring band strength in an image of a gel). Northern blotting is, however, still widely used as the additional mRNA size information allows the discrimination of alternately spliced transcripts.

A more modern low-throughput approach for measuring mRNA abundance is reverse transcription quantitative polymerase chain reaction (RT-PCR followed with qPCR). RT-PCR first generates a DNA template from the mRNA by reverse transcription, which is called cDNA. This cDNA template is then used for qPCR where the change in fluorescence of a probe changes as the DNA amplification process progresses. With a carefully constructed standard curve qPCR can produce an absolute measurement such as number of copies of mRNA, typically in units of copies per nanolitre of homogenized tissue or copies per cell. qPCR is very sensitive (detection of a single mRNA molecule is possible), but can be expensive due to the fluorescent probes required.

An even more advanced approach is to individually tag single mRNA molecules with fluorescent barcodes (nanostrings), which can be detected one-by-one and counted for direct digital quantification. The advantage of this approach is that it does not rely on analog quantification of fluorescent intensity, which can be problematic due to noise, lack of linearity, and narrow dynamic range. Instead, the technique relies on fluorescence to detect simply the presence of a single mRNA molecule in a binary ("yes" or "no") mode. This method was invented by Dr. Krassen Dimitrov at the Institute for Systems Biology and commercialized through his start-up company, NanoString Technologies

Northern blots and RT-qPCR are good for detecting whether a single gene is being expressed, but it quickly becomes impractical if many genes within the sample are being studied. Using DNA microarrays transcript levels for many genes at once (expression profiling) can be measured. Recent advances in microarray technology allow for the quantification, on a single array, of transcript levels for every known gene in several organism's genomes, including humans.

Alternatively "tag based" technologies like Serial analysis of gene expression (SAGE), which can provide a relative measure of the cellular concentration of different mRNAs, can be used. The great advantage of tag-based methods is the "open architecture", allowing for the exact measurement of any transcript, with a known or unknown sequence.

Protein Quantification

For genes encoding proteins the expression level can be directly assessed by a number of means with some clear analogies to the techniques for mRNA quantification.

The most commonly used method is to perform a Western blot against the protein of interest - this gives information on the size of the protein in addition to its identity. A sample (often cellular lysate) is separated on a polyacrylamide gel, transferred to a membrane and then probed with an antibody to the protein of interest. The antibody can either be conjugated to a fluorophore or to horseradish peroxidase for imaging and/or quantification. The gel-based nature of this assay makes quantification less accurate but it has the advantage of being able to identify later modifications to the protein, for example proteolysis or ubiquitination, from changes in size.

Localisation

In situ-hybridization of Drosophila embryos at different developmental stages for the mRNA responsible for the expression of hunchback. High intensity of blue color marks places with high hunchback mRNA quantity.

Analysis of expression is not limited to only quantification; localisation can also be determined. mRNA can be detected with a suitably labelled complementary mRNA strand and protein can be detected via labelled antibodies. The probed sample is then observed by microscopy to identify where the mRNA or protein is. The three-dimensional structure of green fluorescent protein. The residues in the centre of the "barrel" are responsible for production of green light after exposing to higher energetic blue light. From PDB 1EMA.

By replacing the gene with a new version fused a green fluorescent protein (or similar) marker expression may be directly quantified in live cells. This is done by imaging using a fluorescence microscope.

It is very difficult to clone a GFP-fused protein into its native location in the genome without affecting expression levels so this method often cannot be used to measure endogenous gene expression. It is, however, widely used to measure the expression of a gene artificially introduced into the cell, for example via an expression vector. It is important to note that by fusing a target protein to a fluorescent reporter the protein's behavior, including its cellular localization and expression level, can be significantly changed.

The enzyme-linked immunosorbent assay works by using antibodies immobilised on a microtiter plate to capture proteins of interest from samples added to the well. Using a detection antibody conjugated to an enzyme or fluorophore the quantity of bound protein can be accurately measured by fluorometric or colourimetric detection. The detection process is very similar to that of a Western blot, but by avoiding the gel steps more accurate quantification can be achieved.

Expression System

An expression system is a system specifically designed for the production of a gene product of choice. This is normally a protein although may also be RNA, such as tRNA or a ribozyme. An expression system consists of a gene, normally encoded by DNA, and the molecular machinery required to transcribe the DNA into mRNA and translate the mRNA into protein using the reagents provided. In the broadest sense this includes every living cell but the term is more normally used to refer to expression as a laboratory tool. An expression system is therefore often artificial in some manner. Expression systems are, however, a fundamentally natural process. Viruses are an excellent example where they replicate by using the host cell as an expression system for the viral proteins and genome.

In Nature

In addition to these biological tools, certain naturally observed configurations of DNA (genes, promoters, enhancers, repressors) and the associated machinery itself are referred to as an expression system. This term is normally used in the case where a gene or set of genes is switched on under well defined conditions. For example the simple repressor switch expression system in Lambda phage and the lac operator system in bacteria. Several natural expression systems are directly used or modified and used for artificial expression systems such as the Tet-on and Tet-off expression system.

Gene Networks

Genes have sometimes been regarded as nodes in a network, with inputs being proteins such as transcription factors, and outputs being the level of gene expression. The node itself performs a function, and the operation of these functions have been interpreted as performing a kind of information processing within cell and determine cellular behavior. Gene networks can also be constructed without formulating an explicit causal model. This is often the case when assembling networks from large expression data sets. Covariation and correlation of expression is computed across a large sample of cases and measurements (often transcriptome or proteome data). The source of variation can be either experimental or natural (observational). There are several ways to construct gene expression networks, but one common approach is to compute a matrix of all pair-wise correlations of expression across conditions, time points, or individuals and convert the matrix (after thresholding at some cut-off value) into a graphical representation in which nodes represent genes, transcripts, or proteins and edges connecting these nodes represent the strength of association.

Techniques and Tools

The following experimental techniques are used to measure gene expression and are listed in roughly chronological order, starting with the older, more established technologies. They are divided into two groups based on their degree of multiplexity.

- Low-to-mid-plex techniques:
  - o Reporter gene
  - o Northern blot
  - o Western blot
  - o Fluorescent in situ hybridization
  - o Reverse transcription PCR
  - o Digital counting of single transcript molecules.
- Higher-plex techniques:
  - o SAGE
  - o DNA microarray
  - o Tiling array
  - o RNA-Seq.

Transcriptional Bursting

Transcriptional bursting, also known as transcriptional pulsing, is a fundamental property of genes from bacteria to humans. Transcription of genes, the process which transforms the stable code written in DNA into the mobile RNA message can occur in "bursts" or "pulses". This phenomenon has recently come to light with the advent of new technologies, such as MS2 tagging, to detect RNA production in single cells, allowing precise measurements of RNA number, or RNA appearance at the gene. Other, more widespread techniques, such as Northern Blotting, Microarrays, RT-PCR and RNA-Seq, measure bulk RNA levels from homogenous population extracts. These techniques lose dynamic information from individual cells, and give the impression transcription is a continuous smooth process. The reality is that transcription is irregular, with strong periods of activity, interspersed by long periods of inactivity. Averaged over millions of cells, this appears continuous. But at the individual cell level, there is considerable variability, and for most genes, very little activity at any one time.

Bursting may result from the stochastic nature of biochemical events superimposed upon a 2 or more step fluctuation. In its most simple form, the gene can exist in 2 states, one where activity is negligible and one where there is a certain probability of activation. Only in the second state does transcription readily occur.

Whilst the nuclear and signaling landscapes of complex eukaryotic nuclei are likely to favour more than two simple states- for example, there are over twenty post-translational modifications of nucleosomes known, this simple two step model perhaps provides a reasonable intellectual framework for understanding the changing probabilities affecting transcription. It seems likely that some rudimentary eukaryotes have genes which do not show bursting. The genes are always in the permissive state, with a simple probability describing the numbers of RNAs generated.

What do the repressive and permissive states represent? An attractive idea is that the repressed state is a closed chromatin conformation whilst the permissive state is an open one. Another hypothesis is that the fluctuations reflect transition between bound pre-initiation complexes (permissive) and dissociated ones (restrictive). Bursts may also result from bursty signalling, cell cycle effects or movement of chromatin to and from transcription factories.

The bursting phenomenon, as opposed to simple probabilistic models of transcription, can account for the high variability (transcriptional noise) in gene expression occurring between cells in isogenic populations. This variability in turn can have tremendous consequences on cell behaviour, and must be mitigated or integrated.

In certain contexts, such as the survival of microbes in rapidly changing stressful environments, or several types of scattered differentiation, the variability may be essential. Variability also impacts upon the effectiveness of clinical treatment, with resistance of bacteria to antibiotics demonstrably caused by non-genetic differences. Variability in gene expression may also contribute to resistance of sub-populations of cancer cells to chemotherapy.

Expression Profiling

In the field of molecular biology, gene expression profiling is the measurement of the activity (the expression) of thousands of genes at once, to create a global picture of cellular function. These profiles can, for example, distinguish between cells that are actively dividing, or show how the cells react to a particular treatment. Many experiments of this sort measure an entire genome simultaneously, that is, every gene present in a particular cell.

DNA Microarray technology measures the relative activity of previously identified target genes. Sequence based techniques, like serial analysis of gene expression (SAGE, SuperSAGE) are also used for gene expression profiling. SuperSAGE is especially accurate and can measure any active gene, not just a predefined set. The advent of next-generation sequencing has made sequence based expression analysis an increasingly popular, "digital" alternative to microarrays. However, microarrays are far more common, accounting for 17,000 PubMed articles by 2006.

Background

Expression profiling is a logical next step after sequencing a genome: the sequence tells us what the cell could possibly do, while the expression profile tells us what it is actually doing now. Genes contain the instructions for making messenger RNA (mRNA), but at any moment each cell makes mRNA from only a fraction of the genes it carries. If a gene is used to produce mRNA, it is considered "on", otherwise "off". Many factors determine whether a gene is on or off, such as the time of day, whether or not the cell is actively dividing, its local environment, and chemical signals from other cells. Skin cells, liver cells and nerve cells turn on (express) somewhat different genes and that is in large part what makes them different. Therefore, an expression profile allows one to deduce a cell's type, state, environment, and so forth.

Expression profiling experiments often involve measuring the relative amount of mRNA expressed in two or more experimental conditions. This is because altered levels of a specific sequence of mRNA suggest a changed need for the protein coded for by the mRNA, perhaps indicating a homeostatic response or a pathological condition. For example, higher levels of mRNA coding for alcohol dehydrogenase suggest that the cells or tissues under study are responding to increased levels of ethanol in their environment. Similarly, if breast cancer cells express higher levels of mRNA associated with a particular transmembrane receptor than normal cells do, it might be that this receptor plays a role in breast cancer. A drug that interferes with this receptor may prevent or treat breast cancer. In developing a drug, one may perform gene expression profiling experiments to help assess the drug's toxicity, perhaps by looking for changing levels in the expression of cytochrome P450 genes, which may be a biomarker of drug metabolism. Gene expression profiling may become an important diagnostic test.

Comparison to Proteomics

The human genome contains on the order of 25,000 genes which work in concert to produce on the order of 1,000,000 distinct proteins. This is due to alternative splicing, and also because cells make important changes to proteins through posttranslational modification after they first construct them, so a given gene serves as the basis for many possible versions of a particular protein. In any case, a single mass spectrometry experiment can identify about 2,000 proteins or 0.2% of the total. While knowledge of the precise proteins a cell makes (proteomics) is more relevant than knowing how much messenger RNA is made from each gene, gene expression profiling provides the most global picture possible in a single experiment.

Use in Hypothesis Generation and Testing

Sometimes, a scientist already has an idea what is going on, a hypothesis, and he or she performs an expression profiling experiment with the idea of potentially disproving this hypothesis. In other words, the scientist is making a specific prediction about levels of expression that could turn out to be false.

More commonly, expression profiling takes place before enough is known about how genes interact with experimental conditions for a testable hypothesis to exist. With no hypothesis, there is nothing to disprove, but expression profiling can help to identify a candidate hypothesis for future experiments. Most early expression profiling experiments, and many current ones, have this form which is known as class discovery. A popular approach to class discovery involves grouping similar genes or samples together using k-means or hierarchical clustering. The figure above represents the output of a two dimensional cluster, in which similar samples (rows, above) and similar gene probes (columns) were organized so that they would lie close together. The simplest form of class discovery would be to list all the genes that changed by more than a certain amount between two experimental conditions. Class prediction is more difficult than class discovery, but it allows one to answer questions of direct clinical significance such as, given this profile, what is the probability that this patient will respond to this drug? This requires many examples of profiles that responded and did not respond, as well as cross-validation techniques to discriminate between them.

Limitations

In general, expression profiling studies report those genes that showed statistically significant differences under changed experimental conditions. This is typically a small fraction of the genome for several reasons. First, different cells and tissues express a subset of genes as a direct consequence of cellular differentiation so many genes are turned off. Second, many of the genes code for proteins that are required for survival in very specific amounts so many genes do not change. Third, cells use many other mechanisms to regulate proteins in addition to altering the amount of mRNA, so these genes may stay consistently expressed even when protein concentrations are rising and falling. Fourth, financial constraints limit expression profiling experiments to a small number of observations of the same gene under identical conditions, reducing the statistical power of the experiment, making it impossible for the experiment to identify important but subtle changes.

Finally, it takes a great amount of effort to discuss the biological significance of each regulated gene, so scientists often limit their discussion to a subset. Newer microarray analysis techniques automate certain aspects of attaching biological significance to expression profiling results, but this remains a very difficult problem.

The relatively short length of gene lists published from expression profiling experiments limits the extent to which experiments performed in different laboratories appear to agree. Placing expression profiling results in a publicly accessible microarray database makes it possible for researchers to assess expression patterns beyond the scope of published results, perhaps identifying similarity with their own work.

Validation of High Throughput Measurements

Both DNA microarrays and qPCR exploit the preferential binding or "base pairing" of complementary nucleic acid sequences, and both are used in gene expression profiling, often in a serial fashion. While high throughput DNA microarrays lack the quantitative accuracy of qPCR, it takes about the same time to measure the gene expression of a few dozen genes via qPCR as it would to measure an entire genome using DNA microarrays.

So it often makes sense to perform semi-quantitative DNA microarray analysis experiments to identify candidate genes, then perform qPCR on some of the most interesting candidate genes to validate the microarray results. Other experiments, such as a Western blot of some of the protein products of differentially expressed genes, make conclusions based on the expression profile more persuasive, since the mRNA levels do not necessarily correlate to the amount of expressed protein.

Statistical Analysis

Data analysis of microarrays has become an area of intense research. Simply stating that a group of genes were regulated by at least twofold, once a common practice, lacks a solid statistical footing. With five or fewer replicates in each group, typical for microarrays, a single outlier observation can create an apparent difference greater than two-fold. In addition, arbitrarily setting the bar at two-fold is not biologically sound, as it eliminates from consideration many genes with obvious biological significance.

Rather than identify differentially expressed genes using a fold change cutoff, one can use a variety of statistical tests or omnibus tests such as ANOVA, all of which consider both fold change and variability to create a p-value, an estimate of how often we would observe the data by chance alone. Applying p-values to microarrays is complicated by the large number of multiple comparisons (genes) involved. For example, a p-value of .05 is typically thought to indicate significance, since it estimates a 5% probability of observing the data by chance. But with 10,000 genes on a microarray, 500 genes would be identified as significant at $p < .05$ even if there were no difference between the experimental groups.

One obvious solution is to consider significant only those genes meeting a much more stringent p value criterion, e.g., one could perform a Bonferroni correction on the p-values, or use a false discovery rate calculation to adjust p-values in proportion to the number of parallel tests involved. Unfortunately, these approaches may reduce the number of significant genes to zero, even when genes are in fact differentially expressed. Current statistics such as Rank products aim to strike a balance between false discovery of genes due to chance variation and non-discovery of differentially expressed genes. Commonly cited methods include the Significance Analysis of Microarrays (SAM) and a wide variety of methods are available from Bioconductor and a variety of analysis packages from bioinformatics companies.

Selecting a different test usually identifies a different list of significant genes since each test operates under a specific set of assumptions, and places a different emphasis on certain features in the data. Many tests begin with the assumption of a normal distribution in the data, because that seems like a sensible starting point and often produces results that appear more significant.

Some tests consider the joint distribution of all gene observations to estimate general variability in measurements, while others look at each gene in isolation. Many modern microarray analysis techniques involve bootstrapping (statistics), machine learning or Monte Carlo methods.

As the number of replicate measurements in a microarray experiment increases, various statistical approaches yield increasingly similar results, but lack of concordance between different statistical methods makes array results appear less trustworthy. The MAQC Project makes recommendations to guide researchers in selecting more standard methods so that experiments performed in different laboratories will agree better.

Different from the analysis on differentially expressed individual genes, another type of analysis focuses on differential expression or perturbation of pre-defined gene sets and is called gene set analysis. Gene set analysis demonstrated several major advantages over individual gene differential expression analysis. Gene sets are groups of genes that are functionally related according to current knowledge.

Therefore, gene set analysis is considered a knowledge based analysis approach. Commonly used gene sets include those derived from KEGG pathways, Gene Ontology terms, gene groups that share some other functional annotations, such as common transcriptional regulators etc. Representative gene set analysis methods include GSEA, which estimates significance of gene sets based on permutation of sample labels, and GAGE, which tests the significance of gene sets based on permutation of gene labels or a parametric distribution.

Gene Annotation

While the statistics may reliably identify which gene products change under experimental conditions, making biological sense of expression profiling rests on knowing which protein each gene product makes and what function this protein performs. Gene annotation provides functional and other information, for example the location of each gene within a particular chromosome.

Some functional annotations are more reliable than others; some are absent. Gene annotation databases change regularly, and various databases refer to the same protein by different names, reflecting a changing understanding of protein function. Use of standardized gene nomenclature helps address the naming aspect of the problem, but exact matching of transcripts to genes remains an important consideration.

Categorizing Regulated Genes

Having identified some set of regulated genes, the next step in expression profiling involves looking for patterns within the regulated set. Do the proteins made from these genes perform similar functions? Are they chemically similar? Do they reside in similar parts of the cell? Gene ontology analysis provides a standard way to define these relationships. Gene ontologies start with very broad categories, e.g., "metabolic process" and break them down into smaller categories, e.g., "carbohydrate metabolic process" and finally into quite restrictive categories like "inositol and derivative phosphorylation".

Genes have other attributes beside biological function, chemical properties and cellular location. One can compose sets of genes based on proximity to other genes, association with a disease, and relationships with drugs or toxins. The Molecular Signatures Database and the Comparative Toxicogenomics Database are examples of resources to categorize genes in numerous ways.

Finding Patterns among Regulated Genes

Regulated genes are categorized in terms of what they are and what they do, important relationships between genes may emerge. For example, we might see evidence that a certain gene creates a protein to make an enzyme that activates a protein to turn on a second gene on our list. This second gene may be a transcription factor that regulates yet another gene from our list. Observing these links we may begin to suspect that they represent much more than chance associations in the results, and that they are all on our list because of an underlying biological process. On the other hand, it could be that if one selected genes at random, one might find many that seem to have something in common. In this sense, we need rigorous statistical procedures to test whether the emerging biological themes is significant or not. That's where gene set analysis comes in.

Cause and Effect Relationships

Fairly straightforward statistics provide estimates of whether associations between genes on lists are greater than what one would expect by chance. These statistics are interesting, even if they represent a substantial oversimplification of what is really going on. Here is an example. Suppose there are 10,000 genes in an experiment, only 50 (0.5%) of which play a known role in making cholesterol.

The experiment identifies 200 regulated genes. Of those, 40 (20%) turn out to be on a list of cholesterol genes as well. Based on the overall prevalence of the cholesterol genes (0.5%) one expects an average of 1 cholesterol gene for every 200 regulated genes, that is, 0.005 times 200. This expectation is an average, so one expects to see more than one some of the time. The question becomes how often we would see 40 instead of 1 due to pure chance.

According to the hypergeometric distribution one would expect to try about 10^{57} times (10 followed by 56 zeroes) before picking 39 or more of the chlolesterol genes from a pool of 10,000 by drawing 200 genes at random. Whether one pays much attention to how infinitesimally small the probability of observing this by chance is, one would conclude that the regulated gene list is enriched in genes with a known cholesterol association.

One might further hypothesize that the experimental treatment regulates cholesterol, because the treatment seems to selectively regulate genes associated with cholesterol. While this may be true, there are a number of reasons why making this a firm conclusion based on enrichment alone represents an unwarranted leap of faith. One previously mentioned issue has to do with the observation that gene regulation may have no direct impact on protein regulation: even if the proteins coded for by these genes do nothing other than make cholesterol, showing that their mRNA is altered does not directly tell us what is happening at the protein level. It is quite possible that the amount of these cholesterol-related proteins remains constant under the experimental conditions. Second, even if protein levels do change, perhaps there is always enough of them around to make cholesterol as fast as it can be possibly made, that is, another protein, not on our list, is the rate determining step in the process of making cholesterol. Finally, proteins typically play many roles, so these genes may be regulated not because of their shared association with making cholesterol but because of a shared role in a completely independent process.

Bearing the forgoing caveats in mind, while gene profiles do not in themselves prove causal relationships between treatments and biological effects, they do offer unique biological insights that would often be very difficult to arrive at in other ways.

Using Patterns to Find Regulated Genes

As described above, one can identify significantly regulated genes first and then find patterns by comparing the list of significant genes to sets of genes known to share certain associations. One can also work the problem in reverse order. Here is a very simple example. Suppose there are 40 genes associated with a known process, for example, a predisposition to diabetes. Looking at two groups of expression profiles, one for mice fed a high carbohydrate diet and one for mice fed a low carbohydrate diet, one observes that all 40 diabetes genes are expressed at a higher level in the high carbohydrate group than the low carbohydrate group. Regardless of whether any of these genes would have made it to a list of significantly altered genes, observing all 40 up, and none down appears unlikely to be the result of pure chance: flipping 40 heads in a row is predicted to occur about one time in a trillion attempts using a fair coin.

For a type of cell, the group of genes whose combined expression pattern is uniquely characteristic to a given condition constitutes the gene signature of this condition. Ideally, the gene signature can be used to select a group of patients at a specific state of a disease with accuracy that facilitates selection of treatments. Gene Set Enrichment Analysis (GSEA) and similar methods take advantage of this kind of logic but uses more sophisticated statistics, because component genes in real processes display more complex behavior than simply moving up or down as a group, and the amount the genes move up and down is meaningful, not just the direction. In any case, these statistics measure how different the behavior of some small set of genes is compared to genes not in that small set.

GSEA uses a Kolmogorov Smirnov style statistic to see whether any previously defined gene sets exhibited unusual behavior in the current expression profile. This leads to a multiple hypothesis testing challenge, but reasonable methods exist to address it.

Conclusions

Expression profiling provides new information about what genes do under various conditions. Overall, microarray technology produces reliable expression profiles. From this information one can generate new hypotheses about biology or test existing ones. However, the size and complexity of these experiments often results in a wide variety of possible interpretations. In many cases, analyzing expression profiling results takes far more effort than performing the initial experiments.

Most researchers use multiple statistical methods and exploratory data analysis before publishing their expression profiling results, coordinating their efforts with a biostatistician or other expert in microarray technology. Good experimental design, adequate biological replication and follow up experiments play key roles in successful expression profiling experiments.

Paramutation

In epigenetics, paramutation is an interaction between two alleles of a single locus, resulting in a heritable change of one allele that is induced by the other allele. Paramutation violates Mendel's first law, which states that in the process of the formation of the gametes (egg or sperm) the allelic pairs separate, one going to each gamete, and that each allele remains completely uninfluenced by the other.

In paramutation an allele in one generation heritably affects the other allele in future generations, even if the allele causing the change is itself not transmitted. What may be transmitted are patterns of DNA methylation or RNAs such as piRNAs, siRNAs, miRNAs or other regulatory RNAs. Through proper breeding, paramutation can result in isogenic sibling plants with drastically different phenotypes.

Paramutation was first discovered and studied in maize (*Zea mays*) by R.A. Brink at the University of Wisconsin–Madison in the 1950s. Brink noticed that specific weakly expressed alleles of the *red1* (*r1*) locus in maize, which encodes a transcription factor that confers red pigment to corn kernels, can heritably change specific strongly expressed alleles to a weaker expression state. The weaker expression state adopted by the changed allele is heritable and can, in turn, change the expression state of other active alleles in a process termed secondary paramutation. Brink showed that the influence of the paramutagenic allele could persist for many generations.

Interestingly, paramutation can result in a single allele of a gene controlling a spectrum of phenotypes. At *r1* in maize, for example, the weaker expression state adopted by an allele following paramutation can range from completely colorless to nearly fully-colored kernels. This is an exception to the general observation that continuous variation is controlled by many genes.

Allelic interactions similar to paramutation have since been reported in other organisms, including tomato, pea, and mice.

The molecular basis of paramutation is being unraveled, almost exclusively in maize. Paramutation may share common mechanisms with other epigenetic phenomena, such as gene silencing and genomic imprinting. In maize, paramutation seems to share many traits with the well understood RNA-directed DNA-methylation pathway in *Arabidopsis thaliana*, even though it has never been observed in the famous model plant. Alleman (2006) reported that, in maize, "paramutation is RNA-directed. Stability of the chromatin states associated with paramutation and transposon silencing requires the *mop1* gene, which encodes an RNA-dependent RNA polymerase." Exactly how the RNA produced by this polymerase causes paramutation in maize is not yet understood, but like other epigenetic changes, it involves the covalent modification of DNA and/or the DNA-bound histone proteins without changing the DNA sequence itself.

Ridges

Ridges (regions of increased gene expression) are domains of the genome with a high gene expression; the opposite of ridges are antiridges. The term was first used by Caron et al. in 2001. Characteristics of ridges are:

- Gene dense
- Contain much C and G nucleobases
- Genes have short introns
- high SINE repeat density
- low LINE repeat density.

Discovery

Clustering of genes in prokaryotes was known for a long time. Their genes are grouped in operons, genes within operons share a common promoter unit. These genes are mostly functionally related. The genome of prokaryotes is relatively very simple and compact. In eukaryotes the genome is huge and only a small amount of it are functionally genes, furthermore the genes are not arranged in operons. Except for nematodes and trypanosomes; although their operons are different from the prokaryotic operons. In eukaryotes each gene has a transcription regulation site of its own. Therefore genes don't have to be in close proximity to be co-expressed.

Therefore it was long assumed that eukaryotic genes were randomly distributed across the genome due to the high rate of chromosome rearrangements. But because the complete sequence of genomes became available it became possible to absolutely locate a gene and measure its distance to other genes. The first eukaryote genome ever sequenced was that of Saccharomyces cerevisiae, or budding yeast, in 1996. Half a year after that Velculescu et al. (1997) published a research in which they had integrated SAGE data with the now available genome map.

During a cell cycle different genes are active in a cell. Therefore they used SAGE data from three moments of the cell cycle (log phase, S phase-arrested and G2/M-phase arrested cells). Because in yeast all genes have a promoter unit of their own it was not suspected that genes would cluster near to each other but they did. Clusters were present on all 16 yeast chromosomes. A year later Cho et al. also reported (although in more detail) that certain genes are located near to each other in yeast.

Characteristics and Function

Co-expression: Cho et al. were the first who determined that clustered genes have the same expression levels. They identified transcripts that show cell-cycle dependent periodicity. Of those genes 25% was located in close proximity to other genes which were transcript in the same cell cycle. Cohen et al. (2000) also identified clusters of co-expressed genes.

Caron et al. (2001) made a human transcriptome map of 12 different tissues (cancer cells) and concluded that genes are not randomly distributed across the chromosomes. Instead, genes tent to cluster in groups of sometimes 39 genes in close proximity. Clusters were not only gene dense. They identified 27 clusters of genes with very high expression levels and called them RIDGEs. A common RIDGE counts 6 to 30 genes per centiray. However, there were great exceptions, 40 to 50% of the RIDGEs were not that gene dense; just like in yeast these RIDGEs were located in the telomere regions.

Lercher et al. (2002) pointed to some weaknesses in Caron's approach. Clusters of genes in close proximity and high transcription levels can easily been generated by tandem duplicates. Genes can generate duplicates of themselves which are incorporated in their neighborhood. These duplicates can either became a functional part of the pathway of their parent gene, or (because they are no longer favored by natural selection) gain deleterious mutations and turn into pseudogenes. Because these duplicates are false positives in the search for gene clusters they have to be excluded. Lercher excluded neighboring genes with high resemblance to each other, after that he searched with a sliding window for regions with 15 neighboring genes.

It was clear that gene dense regions existed. There was a striking correlation between gene density and a high CG content. Some clusters indeed had high expression levels. But most of the highly expressed regions consisted of housekeeping genes; genes that are highly expressed in all tissues because they code for basal mechanisms. Only a minority of the clusters contained genes that were restricted to specific tissues.

Versteeg et al. (2003) tried, with a better human genome map and better SAGE taqs, to determine the characteristics of RIDGEs more specific. Overlapping genes were treated as one gene, and genes without introns were rejected as pseudogenes. They determined that RIDGEs are very gene dense, have a high gene expression, short introns, high SINE repeat density and low LINE repeat density. Clusters containing genes with very low transcription levels had characteristics that were the opposite of RIDGEs, therefore those clusters were called antiridges. LINE repeats are junk DNA which contains a cleavage site of endonuclease (TTTTA). Their scarcity in RIDGEs can be explained by the fact that natural selection favors the scarcity of LINE repeats in ORFs because their endonuclease sites can cause deleterious mutation to the genes. Why SINE repeats are abundant is not yet understood.

Versteeg et al. also concluded that, contrary to Lerchers analysis, the transcription levels of many genes in RIDGEs (for example a cluster on chromosome 9) can vary strongly between different tissues. Lee et al. (2003) analysed the trend of gene clustering between different species. They compared Saccharomyces cerevisiae, Homo sapiens, Caenorhabditis elegans, Arabidopsis thaliana and Drosophila melanogaster, and found a degree of clustering, as fraction of genes in loose clusters, of respectively (37%), (50%), (74%), (52%) and (68%). They concluded that pathways of which the genes are clusters across many species are rare. They found seven universally clustered pathways: glycolysis, aminoacyl-tRNA biosynthesis, ATP synthase, DNA polymerase, hexachlorocyclohexane degradation, cyanoamino acid metabolism, and photosynthesis (ATP synthesis in non plant species). Not surprisingly these are basic cellular pathways.

Lee et al. used very diverse groups of animals. Within these groups clustering is conserved, for example the clustering motifs of Homo sapiens and Mus musculus are more or less the same.

Spellman and Rubin (2002) made a transcriptome map of Drosophila. Of all assayed genes 20% was clustered. Clusters consisted of 10 to 30 genes over a group size of about 100 kilobases. The members of the clusters were not functionally related and the location of clusters didn't correlate with know chromatin structures.

This study also showed that within clusters the expression levels of on average 15 genes was much the same across the many experimental conditions which were used. These similarities were so striking that the authors reasoned that the genes in the clusters are not individually regulated by their personal promoter but that changes in the chromatin structure were involved. A similar co-regulation pattern was published in the same year by Roy et al. (2002) in C. elegans. Many genes which are grouped into clusters show the same expression profiles in human invasive ductal breast carcinomas. Roughly 20 % of the genes show a correlation with their neighbors. Clusters of co-expressed genes were divided by regions with less correlation between genes. These clusters could cover an entire chromosome arm.

Contrary to previous discussed reports Johnidis et al. (2005) have discovered that (at least some) genes within clusters are not co-regulated. Aire is a transcription factor which has a up- and down-regulation effect on various genes. It functions in negative selection of thymocytes, which responds to the organisms own epitopes, by medullary cells.

The genes that were controlled by aire clustered. 53 of the genes most activated by aire had an aire-activated neighbor within 200 Kb or less, and 32 of the genes most repressed by aire had an aire-repressed neighbor within 200 Kb; this is less than expected by change. They did the same screening for the transcriptional regulator CIITA.

These transcription regulators didn't have the same effect on al genes in the same cluster. Genes that were activated and repressed or unaffected were sometimes present in the same cluster. In this case, it's impossible that aire-regulated genes were clustered because they were all co-regulated.

So it is not very clear if domains are co-regulated or not. A very effective way to test this would be by insert synthetic genes into RIDGEs, antiridges and/or random places in the genome and determine their expression. Those expression levels must be compared to each other. Gierman et al. (2007) were the first who proved co-regulation using this approach. As a insertion construct they used a fluorescing GFP gene driven by the ubiquitously expressed human phosphoglycerate kinase (PGK) promoter. They integrated this construct in 90 different positions in the genome of human HEK293 cells. They found that the expression of the construct in Ridges was indeed higher than those inserted in antiridges (while all constructs have the same promoter).

They investigated if these differences in expressions were due to genes in the direct neighborhood of the constructs or by the domain as a whole. They found that constructs next to highly expressed genes were slightly more expressed than others. But when to enlarged the window size to the surrounding 49 genes (domain level) they saw that constructs located in domains with an overall high expression had a more than 2-fold higher expression then those located in domains with a low expression level.

They also checked if the construct was expressed at similar levels as neighboring genes, and if that tight co-expression was present solely within RIDGEs. They found that the expressions were highly correlated within RIDGEs, and almost absent near the end and outside the RIDGEs.

Previous observations and the research of Gierman et al. proved that the activity of a domain has great impact on the expression of the genes located in it. And the genes within a RIDGE are co-expressed. However the constructs used by Gierman et al. were regulated by al fulltime active promoter. The genes of the research of Johnidis et al. were dependent of the present of the aire transcription factor. The strange expression of the aire regulated genes could partly have been caused by differences in expression and conformation of the aire transcription factor itself.

Functional Relation

It was known before the genomic era that clustered genes tend to be functionally related. Abderrahim et al. (1994) had shown that all the genes of the major histocompatibility complex were clustered on the 6p21 chromosome. Roy et al. (2002) showed that in the nematode C. elegans genes that are solely expressed in muscle tissue during the larval stage tent to cluster in small groups of 2-5 genes. They identified 13 clusters.

Yamashita et al. (2004) showed that genes related to specific functions in organs tent to cluster. Six liver related domains contained genes for xenobiotic, lipid and alcohol metabolism. Five colon-related domains had genes for apoptosis, cell proliferation, ion transporter and mucin production. These clusters were very small and expression levels were low. Brain and breast related genes didn't cluster.

This shows that at least some clusters consist of functionally related genes. However, there are great exceptions. Spellman and Rubin have shown that there are clusters of co-expressed genes that are not functionally related. It seems like that clusters appear in very different forms.

Regulation

Cohen et al. found that of a pair of co-expressed genes only one promoter has an Upstream Activating Sequence (UAS) associated with that expression pattern. They suggested that UASs can activate genes that are not in immediate adjacency to them. This explanation could explain the co-expression of small clusters, but many clusters contain to many genes to be regulated by a single UAS.

Chromatin changes are a plausible explanation for the co-regulation seen in clusters. Chromatin consists of the DNA strand and histones that are attached to the DNA. Regions were chromatin is very tightly packed are called heterochromatin. Heterochromatin consists very often of remains of viral genomes, transposons and other junk DNA. Because of tight packing the DNA is almost unreachable for the transcript machinery, covering deleterious DNA with proteins is the way in which the cell can protect itself. Chromatin which consists of functional genes is often an open structure were the DNA is accessible. However, most of the genes are not needed to be expressed all the time.

DNA with genes that aren't needed can be covered with histones. When a gene must be expressed special proteins can alter the chemical that are attached to the histones (histone modifications) that cause the histones to open the structure. When the chromatin of one gene is opened, the chromatin of the adjacent genes is also until this modification meets a boundary element. In that way genes is close proximity are expressed on the same time. So, genes are clustered in "expression hubs". In comparison with this model Gilbert et al. (2004) showed that RIDGEs are mostly present in open chromatin structures.

However Johnidis et al. (2005) have shown that genes in the same cluster can be very differently expressed. How eukaryotic gene regulation, and associated chromatin changes, precisely works is still very unclear and there is no consensus about it. In order to get a clear picture about the mechanism of gene clusters first the workings chromatin and gene regulation needs to be illuminated. Furthermore most papers that identified clusters of co-regulated genes focused on transcription levels whereas few focused on clusters regulated by the same transcription-factors. Johnides et al. discovered strange phenomena when they did.

Origins

The fist models which tried to explain the clustering of genes were, of course, focused on operons because they were discovered before eukaryote gene clusters were. In 1999 Lawrence proposed a model for the origin operons. This selfish operon model suggests that individual genes were grouped together by vertical en horizontal transfer and were preserved as a single unit because that was beneficial for the genes, not per see for the organism. This model predicts that the gene clusters must have conserved between species. This is not the case for many operons and gene clusters seen in eukaryotes.

According to Eichler and Sankoff the two mean processes in eukaryotic chromosome evolution are 1) rearrangements of chromosomal segments and 2) localized duplication of genes. Clustering could be explained by reasoning that all genes in a cluster are originated from tandem duplicates of a common ancestor. If all co-expressed genes in a cluster were evolved from a common ancestral gene it would have been expected that they're co-expressed because they all have comparable promoters. However, gene clustering is a very common tread in genomes and it isn't clear how this duplication model could explain all of the clustering. Furthermore many genes that are present in clusters are not homologous.

How did evolutionary non-related genes come in close proximity in the first place? Either there is a force that brings functionally related genes near to each other, or the genes came near by change. Singer et al. proposed that genes came in close proximity by random recombination of genome segments. When functionally related genes came in close proximity to each other, this proximity was conserved. They determined all possible recombination sites between genes of human and mouse. After that, they compared the clustering of the mouse and human genome and looked if recombination had occurred at the potentially recombination sites. It turned out that recombination between genes of the same cluster was very rare. So, as soon as a functional cluster is formed recombination is suppressed by the cell. On sex chromosomes, the amount of clusters is very low in both human and mouse. The authors reasoned this was due to the low rate of chromosomal rearrangements of sex chromosomes.

Open chromatin regions are active regions. It is more likely that genes will be transferred to these regions. Genes from organelle and virus genome are inserted more often in these regions. In this way non-homologous genes can be pressed together in a small domain.

It is possible that some regions in the genome are better suited for important genes. It is important for the cell that genes that are responsible for basal functions are protected from recombination. It has been observed in yeast and worms that essential genes tent to cluster in regions with a small replication rate. It is possible that genes came in close proximity by change. Other models have been proposed but none of them can explain all observed phenomena. It's clear that as soon as clusters are formed they are conserved by natural selection. However, a precise model of how genes came in close proximity is still lacking.

The bulk of the present clusters must have formed relatively recent because only seven clusters of functionally related genes are conserved between phyla. Some of these differences can be explained by the fact that gene expression is very differently regulated by different phyla. For example, in vertebrates and plants DNA methylation is used, whereas it is absent in yeast and flies.

AlloMap Molecular Expression Testing

AlloMap molecular expression testing, developed and commercialized by XDx, is a gene expression profiling test to identify heart transplant recipients with a low probability of one type of transplant rejection. The test is performed on a blood sample, providing a non-invasive test to help manage the care of patients post transplant. Prior to the availability of this test, the primary method for managing heart transplant rejection was the invasive technique of endomyocardial biopsy. Test results are reported as a single score indicating the probability of moderate/severe acute cellular rejection (ACR). The performance characteristics of the test make it best suited to help indicate that acute cellar rejection is not present. The score is based on the amount of RNA from each gene in a 20-gene panel comprising 11 rejection-related genes and 9 genes used for normalization and quality control.

Many of the rejection-related genes are associated with biological pathways involved in the immune response and rejection processes. The test score is used, along with other standard clinical assessments, to evaluate the patient's probability of acute cellular rejection and the need for additional evaluations. This test is not designed to be informative about other forms of heart rejections such as antibody-mediated rejection (AMR) or cardiac allograft vasculopathy (CAV).

AlloMap has been commercially available since 2005 as a CLIA approved Laboratory Developed Test (LDT) and was cleared by the U.S. Food and Drug Administration (FDA) in 2008 as a Class II Medical Device. It is available only from the XDx Reference Laboratory in Brisbane, CA. The use of the test is described in the recommendations for the non-invasive monitoring of acute heart transplant rejection in the first evidence-based clinical practice guidelines for the care of heart transplant recipients issued by the International Society of Heart and Lung Transplantation.

Development

The test was developed using genomics and bioinformatics technologies. DNA microarrays were used to discover 252 candidate genes for which the amount of RNA in blood samples was related to rejection. Quantitative real-time polymerase chain reaction technology (qRT-PCR) confirmed 68 of the candidate genes from which the 20-gene gene expression panel was selected. The diagnostic performance was verified using independent patient samples from a multicenter clinical study. Initial clinical experience at three medical centers was published in 2006, confirming the efficacy and performance of the AlloMap test.

Clinical Studies

CARGO Study: The development and clinical validation of the test used patient samples and clinical data obtained during the Cardiac Allograft Rejection Gene Expression Observational (CARGO) Study. From 2001 to 2005, 737 patients from nine U.S. transplant centers enrolled in the Study and contributed 5,834 blood samples and associated clinical data. Initial clinical experience at three medical centers was published in 2006, confirming the efficacy and performance of the test.

IMAGE Study

A comparative effectiveness study, the Invasive Monitoring Attenuation through Gene Expression (IMAGE) Study, compared clinical outcomes of patients managed with AlloMap to outcomes of patients managed with endomyocardial biopsy. The study, which ran from 2005–09, included 602 patients from thirteen U.S. centers who were at least six months post-transplant. The results showed that AlloMap was not inferior to endomyocardial biopsy with respect to clinical outcomes when used to monitor stable, asymptomatic heart transplant patients.

Indications for Use

The test is currently indicated for use in heart transplant recipients 15 years of age or older, and at least 2 months (e"55 days) post-transplant.

Method of Use

The test is based on standard quantitative real-time polymerase chain reaction technology (qRT-PCR) using RNA isolated from peripheral blood mononuclear cells (PBMC). A blood sample is collected, PBMC are isolated, lysed and the released RNA stabilized and frozen (PBMC lysate). RNA is then purified from the PBMC lysate, converted into complementary DNA (cDNA), and mixed with gene-specific primers and probes. The expression of each gene is measured by amplification and fluorescence detection using a qRT-PCR instrument. A mathematical classifier combines the measured expression values for each gene into a single value reported as a score between 0 and 40.

Each score is associated with a negative predictive value (NPV) and a positive predictive value (PPV). The test is characterized by high negative predictive values and is therefore a test used to help identify patients at low probability of rejection. The test has a relatively low positive predictive value, meaning that even when the score is relatively high, the risk of rejection may still be low.

5

Organisation of Genome

Genome organization refers to the sequential, not the structural organization of the genome. Besides the coding exons, the non-coding DNA in Eukaryotes may fall in the following classes:

- Introns. They are DNA sequences inserted between the exons and found in the ORF. They are spliced after the first level of transcription. Most introns are junk inserted within genes.
- Pseudogenes. 'Dead', non-functional copies of genes present elsewhere in the genome, but no longer of any use.
- Retropseudogenes. Like pseudogenes, but have been processed, i.e. lack introns. Produced by the action of reverse transcriptase (RT) on mRNA, and subsequent incorporation of the cDNA into the genome.
- Transposons. Jumping genes, which splice themselves in and out of the genome (in DNA form) randomly, by the action of transposase.
- Retrotransposons. Transcribed into an mRNA, which encodes an RT enzyme, which then copies the mRNA back to DNA and incorporates it into the genome.

Infact in humans only 1.5% of the entire genome length corresponds to coding DNA. This 1.5% codes for about 27,000 genes which in turn code for proteins that are responsible for all the cellular processes.

Genome Analysis?

Genome analysis entails the prediction of genes in uncharacterized genomic sequences. The 21st century has seen the announcement of the draft version of the human genome sequence. Model organisms have been sequenced in both the plant and animal kingdoms.

However, the pace of genome annotation is not matching the pace of genome sequencing. Experimental genome annotation is slow and time consuming. The demand is to be able to develop computational tools for gene prediction.

Computational Gene prediction is relatively simple for the prokaryotes where all the genes are converted into the corresponding mRNA and then into proteins. The process is more complex for eukaryotic cells where the coding DNA sequence is interrupted by random sequences called introns.

Some of the questions which biologists want to answer today are:

• Given a DNA sequence, what part of it codes for a protein and what part of it is junk DNA.

• Classify the junk DNA as intron, untranslated region, transposons, dead genes, regulatory elements etc.

• Divide a newly sequenced genome into the genes (coding) and the non-coding regions.

The importance of genome analysis can be understood by comparing the human and chimpanzee genomes. The chimp and human genomes vary by an average of just 2% i.e. just about 160 enzymes.

A complete genome analysis of the two genomes would give a strong insight into the various mechanisms responsible for the differences.

Arabidopsis and Humans have the same number of genes, though the Arabidopsis genome is around 250 times smaller than humans. How is that ?

The human genome has a lot of junk DNA, specifically transposons and mobile genetic elements. This increases the size of the human genome, though the number of genes is only 27,000.

However, the number of protein products in humans is significantly higher. Many of the sequenced human genes have alternative splice products. In addition, several other processes (e.g. signal transduction) proceed via further protein modifications, such as Glycosylation. Therefore, the number of human protein products could far exceed the number of genes.

Why do plants have such bulky genomes when they are not as complex as some of the higher eukaryotes ?

This is mostly due to two factors: the ability of plants to duplicate their genomes in order to reproduce (a process known as polyploidization) and the susceptibility of plants to mobile genetic elements.

- Polyploidization allows plants to more easily form hybrids when pollen and ova from different species come together. The result of such hybridization events are plants with genomes that are the sum of the two parent genome sizes (as opposed to half of one parent's genome and half of the other parent's genome as in normal sexual reproduction.

- Also, in case of plants, it is fairly common to observe insertion of transposable elements in intergenic regions. This also explains the difference in the sizes of plant genomes among themselves as well.

What is Open Reading Frame (ORF) ?

The region of the nucleotide sequences from the start codon (ATG) to the stop codon is called the Open Reading frame.

Gene finding in organism specially prokaryotes starts form searching for an open reading frames (ORF). An ORF is a sequence of DNA that starts with start codon "ATG" (not always) and ends with any of the three termination codons (TAA, TAG, TGA). Depending on the starting point, there are six possible ways (three on forward strand and three on complementary strand) of translating any nucleotide sequence into amino acid sequence according to the genetic code.These are called reading frames.

While eukaryotic gene finding is altogether a different task as the eukaryotic genes are not continuous and interrupted by intervening noncoding sequences called 'introns'. Moreover organization of genetic information in eukaryotes and prokaryotes is different.

What is Coding Sequence(CDS)? How is it Different from the ORF?

The Coding Sequence (CDS) is the actual region of DNA that is translated to form proteins. While the ORF may contain introns as well, the CDS refers to those nucleotides(concatenated exons) that can be divided into codons which are actually translated into amino acids by the ribosomal translation machinery. In Prokaryotes the ORF and the CDS are the same.

DNA Replication

Before a cell divides, its DNA is replicated (duplicated.) Because the two strands of a DNA molecule have complementary base pairs, the nucleotide sequence of each strand automatically supplies the information needed to produce its partner.

If the two strands of a DNA molecule are separated, each can be used as a pattern or template to produce a complementary strand. Each template and its new complement together then form a new DNA double helix, identical to the original.

Before replication can occur, the length of the DNA double helix about to be copied must be unwound.

In addition, the two strands must be separated, much like the two sides of a zipper, by breaking the weak hydrogen bonds that link the paired bases. Once the DNA strands have been unwound, they must be held apart to expose the bases so that new nucleotide partners can hydrogen-bond to them.

The enzyme DNA polymerase then moves along the exposed DNA strand, joining newly arrived nucleotides into a new DNA strand that is complementary to the template.

Each cell contains a family of more than thirty enzymes to insure the accurate replication of DNA.

Primers

Though DNA polymerase can elongate a polynucleotide strand by adding new nucleotides, it cannot start a strand from scratch because it can only bond new nucleotides to a free sugar (3') end of a nucleotide chain. DNA polymerase requires the assistance of a primer, a previously existing short strand of DNA (or RNA) that is complementary to the first part of the DNA segment being copied. This small strand of nucleotides anneals (binds) by complementary base pairing to the beginning of the area being copied. With the primer in place, DNA polymerase is then able to continue adding the rest of the pairs of the segment until a new double strand of DNA is completed. Primers are formed from free nucleotides in the cell by enzymes called DNA primases.

Replication Occurs Differently on Antiparallel Strands of DNA

That nucleotides can be added only to the sugar or 3' end of the growing complementary chain presents no problem for the side of the DNA chain opening at its phosphate or 5' end. The primer that binds to the first few exposed bases will end with a sugar (3') where the phosphate of a new nucleotide can be attached. From there on, DNA polymerase can continuously synthesize the growing complementary strand. This strand of DNA is called the leading strand.

A different challenge faces DNA polymerase when the complementary side of the DNA molecule begins unzipping from its sugar (3') toward its phosphate (5') end. A primer of complementary molecules attaching to the opening end of this chain would have a phosphate not a sugar at its exposed end so that new nucleotides could not be joined.

To get around this problem, this strand is synthesized in small pieces backward from the overall direction of replication. This strand is called the lagging strand. The short segments of newly assembled DNA from which the lagging strand is built are called Okazaki fragments. As replication proceeds and nucleotides are added to the 3' end of the Okazaki fragments, they come to meet each other. The primer fragments are then booted out by enzymes and replaced by appropriate DNA nucleotides. The whole thing is then stitched together by another enzyme called DNA ligase.

DNA Damage, Repair, and Recombination

DNA repair refers to a collection of processes by which a cell identifies and corrects damage to the DNA molecules that encode its genome. In human cells, both normal metabolic activities and environmental factors such as UV light and radiation can cause DNA damage, resulting in as many as 1 million individual molecular lesions per cell per day.

Many of these lesions cause structural damage to the DNA molecule and can alter or eliminate the cell's ability to transcribe the gene that the affected DNA encodes. Other lesions induce potentially harmful mutations in the cell's genome, which affect the survival of its daughter cells after it undergoes mitosis. As a consequence, the DNA repair process is constantly active as it responds to damage in the DNA structure. When normal repair processes fail, and when cellular apoptosis does not occur, irreparable DNA damage may occur, including double-strand breaks and DNA crosslinkages.

The rate of DNA repair is dependent on many factors, including the cell type, the age of the cell, and the extracellular environment. A cell that has accumulated a large amount of DNA damage, or one that no longer effectively repairs damage incurred to its DNA, can enter one of three possible states:

1. an irreversible state of dormancy, known as senescence
2. cell suicide, also known as apoptosis or programmed cell death
3. unregulated cell division, which can lead to the formation of a tumor that is cancerous.

The DNA repair ability of a cell is vital to the integrity of its genome and thus to its normal functioning and that of the organism. Many genes that were initially shown to influence life span have turned out to be involved in DNA damage repair and protection. Failure to correct molecular lesions in cells that form gametes can introduce mutations into the genomes of the offspring and thus influence the rate of evolution.

DNA Damage

DNA damage, due to environmental factors and normal metabolic processes inside the cell, occurs at a rate of 1,000 to 1,000,000 molecular lesions per cell per day. While this constitutes only 0.000165% of the human genome's approximately 6 billion bases (3 billion base pairs), unrepaired lesions in critical genes (such as tumor suppressor genes) can impede a cell's ability to carry out its function and appreciably increase the likelihood of tumour formation.

The vast majority of DNA damage affects the primary structure of the double helix; that is, the bases themselves are chemically modified. These modifications can in turn disrupt the molecules' regular helical structure by introducing non-native chemical bonds or bulky adducts that do not fit in the standard double helix. Unlike proteins and RNA, DNA usually lacks tertiary structure and therefore damage or disturbance does not occur at that level. DNA is, however, supercoiled and wound around "packaging" proteins called histones (in eukaryotes), and both superstructures are vulnerable to the effects of DNA damage.

Sources of Damage

DNA damage can be subdivided into two main types:

1. endogenous damage such as attack by reactive oxygen species produced from normal metabolic byproducts (spontaneous mutation), especially the process of oxidative deamination
   - also includes replication errors.
2. exogenous damage caused by external agents such as
   - ultraviolet [UV 200-300nm] radiation from the sun

- other radiation frequencies, including x-rays and gamma rays
- hydrolysis or thermal disruption
- certain plant toxins
- human-made mutagenic chemicals, especially aromatic compounds that act as DNA intercalating agents
- cancer chemotherapy and radiotherapy
- viruses.

The replication of damaged DNA before cell division can lead to the incorporation of wrong bases opposite damaged ones. Daughter cells that inherit these wrong bases carry mutations from which the original DNA sequence is unrecoverable (except in the rare case of a back mutation, for example, through gene conversion).

Types of Damage

There are five main types of damage to DNA due to endogenous cellular processes:

1. *oxidation* of bases [e.g. 8-oxo-7,8-dihydroguanine (8-oxoG)] and generation of DNA strand interruptions from reactive oxygen species,

2. *alkylation* of bases (usually methylation), such as formation of 7-methylguanine, 1-methyladenine, 6-O-Methylguanine

3. *hydrolysis* of bases, such as deamination, depurination, and depyrimidination.

4. "bulky adduct formation" (i.e., benzo[a]pyrene diol epoxide-dG adduct)

5. *mismatch* of bases, due to errors in DNA replication, in which the wrong DNA base is stitched into place in a newly forming DNA strand, or a DNA base is skipped over or mistakenly inserted.

Damage caused by exogenous agents comes in many forms. Some examples are:

1. *UV-B light* causes crosslinking between adjacent cytosine and thymine bases creating *pyrimidine dimers*. This is called direct DNA damage.

2. *UV-A light* creates mostly free radicals. The damage caused by free radicals is called indirect DNA damage.

3. *Ionizing radiation* such as that created by radioactive decay or in *cosmic rays* causes breaks in DNA strands. Low-level ionizing radiation may induce irreparable DNA damage (leading to replicational and transcriptional errors needed for neoplasia or may trigger viral interactions) leading to pre-mature aging and cancer.

4. *Thermal disruption* at elevated temperature increases the rate of depurination (loss of purine bases from the DNA backbone) and single-strand breaks. For example, hydrolytic depurination is seen in the thermophilic bacteria, which grow in hot springs at 40-80 °C. The rate of depurination (300 purine residues per genome per generation) is too high in these species to be repaired by normal repair machinery, hence a possibility of an adaptive response cannot be ruled out.

5. *Industrial chemicals* such as vinyl chloride and hydrogen peroxide, and environmental chemicals such as polycyclic hydrocarbons found in smoke, soot and tar create a huge diversity of DNA adducts- ethenobases, oxidized bases, alkylated phosphotriesters and Crosslinking of DNA just to name a few.

UV damage, alkylation/methylation, X-ray damage and oxidative damage are examples of induced damage. Spontaneous damage can include the loss of a base, deamination, sugar ring puckering and tautomeric shift.

Nuclear versus Mitochondrial DNA Damage

In human cells, and eukaryotic cells in general, DNA is found in two cellular locations - inside the nucleus and inside the mitochondria. Nuclear DNA (nDNA) exists as chromatin during non-replicative stages of the cell cycle and is condensed into aggregate structures known as chromosomes during cell division. In either state the DNA is highly compacted and wound up around bead-like proteins called histones.

Whenever a cell needs to express the genetic information encoded in its nDNA the required chromosomal region is unravelled, genes located therein are expressed, and then the region is condensed back to its resting conformation. Mitochondrial DNA (mtDNA) is located inside mitochondria organelles, exists in multiple copies, and is also tightly associated with a number of proteins to form a complex known as the nucleoid. Inside mitochondria, reactive oxygen species (ROS), or free radicals, byproducts of the constant production of adenosine triphosphate (ATP) via oxidative phosphorylation, create a highly oxidative environment that is known to damage mtDNA. A critical enzyme in counteracting the toxicity of these species is superoxide dismutase, which is present in both the mitochondria and cytoplasm of eukaryotic cells.

Senescence and Apoptosis

Senescence, an irreversible state in which the cell no longer divides, is a protective response to the shortening of the chromosome ends. The telomeres are long regions of repetitive noncoding DNA that cap chromosomes and undergo partial degradation each time a cell undergoes division. In contrast, quiescence is a reversible state of cellular dormancy that is unrelated to genome damage. Senescence in cells may serve as a functional alternative to apoptosis in cases where the physical presence of a cell for spatial reasons is required by the organism, which serves as a "last resort" mechanism to prevent a cell with damaged DNA from replicating inappropriately in the absence of pro-growth cellular signaling. Unregulated cell division can lead to the formation of a tumour, which is potentially lethal to an organism. Therefore, the induction of senescence and apoptosis is considered to be part of a strategy of protection against cancer.

DNA Damage and Mutation

It is important to distinguish between DNA damage and mutation, the two major types of error in DNA. DNA damages and mutation are fundamentally different. Damages are physical abnormalities in the DNA, such as single- and double-strand breaks, 8-hydroxydeoxyguanosine residues, and polycyclic aromatic hydrocarbon adducts. DNA damages can be recognized by enzymes, and, thus, they can be correctly repaired if redundant information, such as the undamaged sequence in the complementary DNA strand or in a homologous chromosome, is available for copying. If a cell retains DNA damage, transcription of a gene can be prevented, and, thus, translation into a protein will also be blocked. Replication may also be blocked and/or the cell may die.

In contrast to DNA damage, a mutation is a change in the base sequence of the DNA. A mutation cannot be recognized by enzymes once the base change is present in both DNA strands, and, thus, a mutation cannot be repaired. At the cellular level, mutations can cause alterations in protein function and regulation. Mutations are replicated when the cell replicates. In a population of cells, mutant cells will increase or decrease in frequency according to the effects of the mutation on the ability of the cell to survive and reproduce. Although distinctly different from each other, DNA damages and mutations are related because DNA damages often cause errors of DNA synthesis during replication or repair; these errors are a major source of mutation.

Given these properties of DNA damage and mutation, it can be seen that DNA damages are a special problem in non-dividing or slowly dividing cells, where unrepaired damages will tend to accumulate over time. On the other hand, in rapidly dividing cells, unrepaired DNA damages that do not kill the cell by blocking replication will tend to cause replication errors and thus mutation. The great majority of mutations that are not neutral in their effect are deleterious to a cell's survival. Thus, in a population of cells comprising a tissue with replicating cells, mutant cells will tend to be lost. However, infrequent mutations that provide a survival advantage will tend to clonally expand at the expense of neighboring cells in the tissue. This advantage to the cell is disadvantageous to the whole organism, because such mutant cells can give rise to cancer. Thus, DNA damages in frequently dividing cells, because they give rise to mutations, are a prominent cause of cancer. In contrast, DNA damages in infrequently dividing cells are likely a prominent cause of aging.

DNA Repair Mechanisms

Cells cannot function if DNA damage corrupts the integrity and accessibility of essential information in the genome (but cells remain superficially functional when so-called "non-essential" genes are missing or damaged). Depending on the type of damage inflicted on the DNA's double helical structure, a variety of repair strategies have evolved to restore lost information. If possible, cells use the unmodified complementary strand of the DNA or the sister chromatid as a template to recover the original information. Without access to a template, cells use an error-prone recovery mechanism known as translesion synthesis as a last resort. Damage to DNA alters the spatial configuration of the helix, and such alterations can be detected by the cell. Once damage is localized, specific DNA repair molecules bind at or near the site of damage, inducing other molecules to bind and form a complex that enables the actual repair to take place.

Direct Reversal

Cells are known to eliminate three types of damage to their DNA by chemically reversing it. These mechanisms do not require a template, since the types of damage they counteract can occur in only one of the four bases. Such direct reversal mechanisms are specific to the type of damage incurred and do not involve breakage of the phosphodiester backbone.

The formation of pyrimidine dimers upon irradiation with UV light results in an abnormal covalent bond between adjacent pyrimidine bases. The photoreactivation process directly reverses this damage by the action of the enzyme photolyase, whose activation is obligately dependent on energy absorbed from blue/UV light (300–500 nm wavelength) to promote catalysis. Another type of damage, methylation of guanine bases, is directly reversed by the protein methyl guanine methyl transferase (MGMT), the bacterial equivalent of which is called ogt. This is an expensive process because each MGMT molecule can be used only once; that is, the reaction is stoichiometric rather than catalytic. A generalized response to methylating agents in bacteria is known as the adaptive response and confers a level of resistance to alkylating agents upon sustained exposure by upregulation of alkylation repair enzymes. The third type of DNA damage reversed by cells is certain methylation of the bases cytosine and adenine.

Single-strand Damage

When only one of the two strands of a double helix has a defect, the other strand can be used as a template to guide the correction of the damaged strand. In order to repair damage to one of the two paired molecules of DNA, there exist a number of excision repair mechanisms that remove the damaged nucleotide and replace it with an undamaged nucleotide complementary to that found in the undamaged DNA strand.

1. Base excision repair (BER), which repairs damage to a single base caused by oxidation, alkylation, hydrolysis, or deamination. The damaged base is removed by a DNA glycosylase. The "missing tooth" is then recognised by an enzyme called AP endonuclease, which cuts the Phosphodiester bond. The missing part is then resynthesized by a DNA polymerase, and a DNA ligase performs the final nick-sealing step.

2. Nucleotide excision repair (NER), which recognizes bulky, helix-distorting lesions such as pyrimidine dimers and 6,4

photoproducts. A specialized form of NER known as transcription-coupled repair deploys NER enzymes to genes that are being actively transcribed.

3. Mismatch repair (MMR), which corrects errors of DNA replication and recombination that result in mispaired (but undamaged) nucleotides.

Double-strand Breaks

Double-strand breaks, in which both strands in the double helix are severed, are particularly hazardous to the cell because they can lead to genome rearrangements. Three mechanisms exist to repair double-strand breaks (DSBs): non-homologous end joining (NHEJ), microhomology-mediated end joining (MMEJ), and homologous recombination. PVN Acharya noted that double-strand breaks and a "cross-linkage joining both strands at the same point is irreparable because neither strand can then serve as a template for repair. The cell will die in the next mitosis or in some rare instances, mutate."

In NHEJ, DNA Ligase IV, a specialized DNA ligase that forms a complex with the cofactor XRCC4, directly joins the two ends. To guide accurate repair, NHEJ relies on short homologous sequences called microhomologies present on the single-stranded tails of the DNA ends to be joined. If these overhangs are compatible, repair is usually accurate. NHEJ can also introduce mutations during repair. Loss of damaged nucleotides at the break site can lead to deletions, and joining of nonmatching termini forms translocations. NHEJ is especially important before the cell has replicated its DNA, since there is no template available for repair by homologous recombination. There are "backup" NHEJ pathways in higher eukaryotes. Besides its role as a genome caretaker, NHEJ is required for joining hairpin-capped double-strand breaks induced during V(D)J recombination, the process that generates diversity in B-cell and T-cell receptors in the vertebrate immune system.

Homologous recombination requires the presence of an identical or nearly identical sequence to be used as a template for repair of the break. The enzymatic machinery responsible for this repair process is nearly identical to the machinery responsible for chromosomal crossover during meiosis. This pathway allows a damaged chromosome to be repaired using a sister chromatid (available in G2 after DNA replication) or a homologous chromosome as a template. DSBs caused by the replication machinery attempting to synthesize across a single-strand break or unrepaired lesion cause collapse of the replication fork and are typically repaired by recombination.

Topoisomerases introduce both single- and double-strand breaks in the course of changing the DNA's state of supercoiling, which is especially common in regions near an open replication fork. Such breaks are not considered DNA damage because they are a natural intermediate in the topoisomerase biochemical mechanism and are immediately repaired by the enzymes that created them.

A team of French researchers bombarded *Deinococcus radiodurans* to study the mechanism of double-strand break DNA repair in that organism. At least two copies of the genome, with random DNA breaks, can form DNA fragments through annealing. Partially overlapping fragments are then used for synthesis of homologous regions through a moving D-loop that can continue extension until they find complementary partner strands. In the final step there is crossover by means of RecA-dependent homologous recombination.

Translesion Synthesis

Translesion synthesis is a DNA damage tolerance process that allows the DNA replication machinery to replicate past DNA lesions such as thymine dimers or AP sites. It involves switching out regular DNA polymerases for specialized translesion polymerases (i.e. DNA polymerase IV or V, from the Y Polymerase family), often with larger active sites that can facilitate the insertion of bases opposite damaged nucleotides. The polymerase switching is thought to be mediated by, among other factors, the post-translational modification of the replication processivity factor PCNA. Translesion synthesis polymerases often have low fidelity (high propensity to insert wrong bases) on undamaged templates relative to regular polymerases. However, many are extremely efficient at inserting correct bases opposite specific types of damage. For example, Pol ç mediates error-free bypass of lesions induced by UV irradiation, whereas Pol æ introduces mutations at these sites. From a cellular perspective, risking the introduction of point mutations during translesion synthesis may be preferable to resorting to more drastic mechanisms of DNA repair, which may cause gross chromosomal aberrations or cell death. In short, the process involves specialized polymerases either bypassing or repairing lesions at locations of stalled DNA replication. A bypass platform is provided to these polymerases by Proliferating cell nuclear antigen (PCNA). Under normal circumstances, PCNA bound to polymerases replicates the DNA. At a site of lesion, PCNA is ubiquitinated, or modified, by the RAD6/RAD18 proteins to provide a platform for the specialized polymerases to bypass the lesion and resume DNA replication.

Global Response to DNA Damage

Cells exposed to ionizing radiation, ultraviolet light or chemicals are prone to acquire multiple sites of bulky DNA lesions and double-strand breaks. Moreover, DNA damaging agents can damage other biomolecules such as proteins, carbohydrates, lipids, and RNA. The accumulation of damage, to be specific, double-strand breaks or adducts stalling the replication forks, are among known stimulation signals for a global response to DNA damage. The global response to damage is an act directed toward the cells' own preservation and triggers multiple pathways of macromolecular repair, lesion bypass, tolerance, or apoptosis. The common features of global response are induction of multiple genes, cell cycle arrest, and inhibition of cell division.

DNA Damage Checkpoints

After DNA damage, cell cycle checkpoints are activated. Checkpoint activation pauses the cell cycle and gives the cell time to repair the damage before continuing to divide. DNA damage checkpoints occur at the G1/S and G2/M boundaries. An intra-S checkpoint also exists. Checkpoint activation is controlled by two master kinases, ATM and ATR. ATM responds to DNA double-strand breaks and disruptions in chromatin structure, whereas ATR primarily responds to stalled replication forks. These kinases phosphorylate downstream targets in a signal transduction cascade, eventually leading to cell cycle arrest. A class of checkpoint mediator proteins including BRCA1, MDC1, and 53BP1 has also been identified. These proteins seem to be required for transmitting the checkpoint activation signal to downstream proteins.

P53 is an important downstream target of ATM and ATR, as it is required for inducing apoptosis following DNA damage. At the G1/S checkpoint, p53 functions by deactivating the CDK2/cyclin E complex. Similarly, p21 mediates the G2/M checkpoint by deactivating the CDK1/cyclin B complex.

The Prokaryotic SOS Response

The SOS response is the term used to describe changes in gene expression in *Escherichia coli* and other bacteria in response to extensive DNA damage. The prokaryotic SOS system is regulated by two key proteins: LexA and RecA. The LexA homodimer is a transcriptional repressor that binds to operator sequences commonly referred to as SOS boxes.

In *Escherichia coli* it is known that LexA regulates transcription of approximately 48 genes including the lexA and recA genes. The SOS response is known to be widespread in the Bacteria domain, but it is mostly absent in some bacterial phyla, like the Spirochetes.

The most common cellular signals activating the SOS response are regions of single-stranded DNA (ssDNA), arising from stalled replication forks or double-strand breaks, which are processed by DNA helicase to separate the two DNA strands. In the initiation step, RecA protein binds to ssDNA in an ATP hydrolysis driven reaction creating RecA–ssDNA filaments. RecA–ssDNA filaments activate LexA autoprotease activity, which ultimately leads to cleavage of LexA dimer and subsequent LexA degradation. The loss of LexA repressor induces transcription of the SOS genes and allows for further signal induction, inhibition of cell division and an increase in levels of proteins responsible for damage processing.

In *Escherichia coli*, SOS boxes are 20-nucleotide long sequences near promoters with palindromic structure and a high degree of sequence conservation. In other classes and phyla, the sequence of SOS boxes varies considerably, with different length and composition, but it is always highly conserved and one of the strongest short signals in the genome. The high information content of SOS boxes permits differential binding of LexA to different promoters and allows for timing of the SOS response. The lesion repair genes are induced at the beginning of SOS response. The error-prone translesion polymerases, for example, UmuCD'2 (also called DNA polymerase V), are induced later on as a last resort. Once the DNA damage is repaired or bypassed using polymerases or through recombination, the amount of single-stranded DNA in cells is decreased, lowering the amounts of RecA filaments decreases cleavage activity of LexA homodimer, which then binds to the SOS boxes near promoters and restores normal gene expression.

Eukaryotic Transcriptional Responses to DNA Damage

Eukaryotic cells exposed to DNA damaging agents also activate important defensive pathways by inducing multiple proteins involved in DNA repair, cell cycle checkpoint control, protein trafficking and degradation. Such genome wide transcriptional response is very complex and tightly regulated, thus allowing coordinated global response to damage. Exposure of yeast *Saccharomyces cerevisiae* to DNA damaging agents results in overlapping but distinct transcriptional profiles. Similarities to environmental shock response indicates that a general global stress response pathway exist at the level of transcriptional activation. In contrast, different human cell types respond to damage differently indicating an absence of a common global response. The probable explanation for this difference between yeast and human cells may be in the heterogeneity of mammalian cells. In an animal different types of cells are distributed among different organs that have evolved different sensitivities to DNA damage.

In general global response to DNA damage involves expression of multiple genes responsible for postreplication repair, homologous recombination, nucleotide excision repair, DNA damage checkpoint, global transcriptional activation, genes controlling mRNA decay, and many others. A large amount of damage to a cell leaves it with an important decision: undergo apoptosis and die, or survive at the cost of living with a modified genome. An increase in tolerance to damage can lead to an increased rate of survival that will allow a greater accumulation of mutations. Yeast Rev1 and human polymerase ç are members of [Y family translesion DNA polymerases present during global response to DNA damage and are responsible for enhanced mutagenesis during a global response to DNA damage in eukaryotes.

DNA Repair and Aging

Pathological Effects of Poor DNA Repair

DNA repair rate is an important determinant of cell pathology.

Experimental animals with genetic deficiencies in DNA repair often show decreased life span and increased cancer incidence. For example, mice deficient in the dominant NHEJ pathway and in telomere maintenance mechanisms get lymphoma and infections more often, and, as a consequence, have shorter lifespans than wild-type mice. In similar manner, mice deficient in a key repair and transcription protein that unwinds DNA helices have premature onset of aging-related diseases and consequent shortening of lifespan. However, not every DNA repair deficiency creates exactly the predicted effects; mice deficient in the NER pathway exhibited shortened life span without correspondingly higher rates of mutation.

If the rate of DNA damage exceeds the capacity of the cell to repair it, the accumulation of errors can overwhelm the cell and result in early senescence, apoptosis, or cancer. Inherited diseases associated with faulty DNA repair functioning result in premature aging, increased sensitivity to carcinogens, and correspondingly increased cancer risk. On the other hand, organisms with enhanced DNA repair systems, such as *Deinococcus radiodurans*, the most radiation-resistant known organism, exhibit remarkable resistance to the double-strand break-inducing effects of radioactivity, likely due to enhanced efficiency of DNA repair and especially NHEJ.

Longevity and Caloric Restriction

A number of individual genes have been identified as influencing variations in life span within a population of organisms. The effects of these genes is strongly dependent on the environment, in particular, on the organism's diet. Caloric restriction reproducibly results in extended lifespan in a variety of organisms, likely via nutrient sensing pathways and decreased metabolic rate. The molecular mechanisms by which such restriction results in lengthened lifespan are as yet unclear; however, the behavior of many genes known to be involved in DNA repair is altered under conditions of caloric restriction.

For example, increasing the gene dosage of the gene SIR-2, which regulates DNA packaging in the nematode worm *Caenorhabditis elegans*, can significantly extend lifespan. The mammalian homolog of SIR-2 is known to induce downstream DNA repair factors involved in NHEJ, an activity that is especially promoted under conditions of caloric restriction. Caloric restriction has been closely linked to the rate of base excision repair in the nuclear DNA of rodents, although similar effects have not been observed in mitochondrial DNA.

It is interesting to note that the *C. elegans* gene AGE-1, an upstream effector of DNA repair pathways, confers dramatically extended life span under free-feeding conditions but leads to a decrease in reproductive fitness under conditions of caloric restriction. This observation supports the pleiotropy theory of the biological origins of aging, which suggests that genes conferring a large survival advantage early in life will be selected for even if they carry a corresponding disadvantage late in life.

Medicine and DNA Repair Modulation

Hereditary DNA Repair Disorders: Defects in the NER mechanism are responsible for several genetic disorders, including:

- Xeroderma pigmentosum: hypersensitivity to sunlight/UV, resulting in increased skin cancer incidence and premature aging
- Cockayne syndrome: hypersensitivity to UV and chemical agents
- Trichothiodystrophy: sensitive skin, brittle hair and nails.

Mental retardation often accompanies the latter two disorders, suggesting increased vulnerability of developmental neurons.

Other DNA repair disorders include:

- Werner's syndrome: premature aging and retarded growth
- Bloom's syndrome: sunlight hypersensitivity, high incidence of malignancies (especially leukemias).
- Ataxia telangiectasia: sensitivity to ionizing radiation and some chemical agents.

All of the above diseases are often called "segmental progerias" ("accelerated aging diseases") because their victims appear elderly and suffer from aging-related diseases at an abnormally young age, while not manifesting all the symptoms of old age.

Other diseases associated with reduced DNA repair function include Fanconi's anemia, hereditary breast cancer and hereditary colon cancer.

DNA Repair and Cancer

Inherited mutations that affect DNA repair genes are strongly associated with high cancer risks in humans. Hereditary nonpolyposis colorectal cancer (HNPCC) is strongly associated with specific mutations in the DNA mismatch repair pathway. BRCA1 and BRCA2, two famous mutations conferring a hugely increased risk of breast cancer on carriers, are both associated with a large number of DNA repair pathways, especially NHEJ and homologous recombination.

Cancer therapy procedures such as chemotherapy and radiotherapy work by overwhelming the capacity of the cell to repair DNA damage, resulting in cell death. Cells that are most rapidly dividing - most typically cancer cells - are preferentially affected. The side-effect is that other non-cancerous but rapidly dividing cells such as stem cells in the bone marrow are also affected. Modern cancer treatments attempt to localize the DNA damage to cells and tissues only associated with cancer, either by physical means (concentrating the therapeutic agent in the region of the tumour) or by biochemical means (exploiting a feature unique to cancer cells in the body).

DNA Repair and Evolution

The basic processes of DNA repair are highly conserved among both prokaryotes and eukaryotes and even among bacteriophage (viruses that infect bacteria); however, more complex organisms with more complex genomes have correspondingly more complex repair mechanisms. The ability of a large number of protein structural motifs to catalyze relevant chemical reactions has played a significant role in the elaboration of repair mechanisms during evolution. For an extremely detailed review of hypotheses relating to the evolution of DNA repair.

The fossil record indicates that single-cell life began to proliferate on the planet at some point during the Precambrian period, although exactly when recognizably modern life first emerged is unclear. Nucleic acids became the sole and universal means of encoding genetic information, requiring DNA repair mechanisms that in their basic form have been inherited by all extant life forms from their common ancestor. The emergence of Earth's oxygen-rich atmosphere (known as the "oxygen catastrophe") due to photosynthetic organisms, as well as the presence of potentially damaging free radicals in the cell due to oxidative phosphorylation, necessitated the evolution of DNA repair mechanisms that act specifically to counter the types of damage induced by oxidative stress.

Rate of Evolutionary Change

On some occasions, DNA damage is not repaired, or is repaired by an error-prone mechanism that results in a change from the original sequence. When this occurs, mutations may propagate into the genomes of the cell's progeny. Should such an event occur in a germ line cell that will eventually produce a gamete, the mutation has the potential to be passed on to the organism's offspring. The rate of evolution in a particular species (or, in a particular gene) is a function of the rate of mutation. As a consequence, the rate and accuracy of DNA repair mechanisms have an influence over the process of evolutionary change.

Senescence

Senescence or biological aging is the change in the biology of an organism as it ages after its maturity. Such changes range from those affecting its cells and their function to that of the whole organism. There are a number of theories as to why senescence occurs, including ones that claim it is programmed by gene expression changes and that it is the accumulative damage of biological processes.

The word *senescence* is derived from the Latin word *senex,* meaning *old man, old age,* or *advanced in age.*

Cellular Senescence

Cellular senescence is the phenomenon by which normal diploid cells lose the ability to divide, normally after about 50 cell divisions in vitro. Some cells become senescent after fewer replications cycles as a result of DNA double strand breaks, toxins, etc. This phenomenon is also known as "replicative senescence", the "Hayflick phenomenon", or the Hayflick limit in honour of Dr. Leonard Hayflick who was the first to publish this information in 1965. In response to DNA damage (including shortened telomeres), cells either age or self-destruct (apoptosis, programmed cell death) if the damage cannot be repaired. In this 'cellular suicide', the death of one cell, or more, may benefit the organism as a whole. For example, in plants the death of the water-conducting xylem cells (tracheids and vessel elements) allows the cells to function more efficiently and so deliver water to the upper parts of a plant.

Aging of the Whole Organism

Organismal senescence is the aging of whole organisms. In general, aging is characterized by the declining ability to respond to stress, increased homeostatic imbalance, and increased risk of aging-associated diseases. Death is the ultimate consequence of aging, though "old age" is not a scientifically recognized cause of death because there is always a specific proximal cause, such as cancer, heart disease, or liver failure. Aging of whole organisms is therefore a complex process that can be defined as "a progressive deterioration of physiological function, an intrinsic age-related process of loss of viability and increase in vulnerability".

Differences in maximum life span among species correspond to different "rates of aging". For example, inherited differences in the rate of aging make a mouse elderly at 3 years and a human elderly at 80 years. These genetic differences affect a variety of physiological processes, including the efficiency of DNA repair, antioxidant enzymes, and rates of free radical production.

Senescence of the organism gives rise to the Gompertz–Makeham law of mortality, which says that mortality rate rises rapidly with age.

Some animals, such as some reptiles and fish, age slowly (negligible senescence) and exhibit very long lifespans. Some even exhibit "negative senescence", in which mortality falls with age, in disagreement with the Gompertz–Makeham "law".

Whether replicative senescence (Hayflick limit) plays a causative role in organismal aging is at present an active area of investigation.

Theories of Aging

The process of senescence is complex, and may derive from a variety of different mechanisms and exist for a variety of different reasons. However, senescence is not universal, and scientific evidence suggests that cellular senescence evolved in certain species because it prevents the onset of cancer. In a few simple species, such as Hydra, senescence is negligible and cannot be detected. All such species have no "post-mitotic" cells; they reduce the effect of damaging free radicals by cell division and dilution. Such species are not immortal, however, as they will eventually fall prey to trauma or disease. Moreover, average lifespans can vary greatly within and between species. This suggests that both genetic and environmental factors contribute to aging.

In general, theories that explain senescence have been divided between the programmed and stochastic theories of aging. Programmed theories imply that aging is regulated by biological clocks operating throughout the lifespan. This regulation would depend on changes in gene expression that affect the systems responsible for maintenance, repair, and defense responses. Stochastic theories blame environmental impacts on living organisms that induce cumulative damage at various levels as the cause of aging, examples of which ranging from damage to DNA, damage to tissues and cells by oxygen radicals (widely known as free radicals countered by the even more well-known antioxidants), and cross-linking.

However, aging is seen as a progressive failure of homeodynamics (homeostasis) involving genes for the maintenance and repair, stochastic events leading to molecular damage and molecular heterogeneity, and chance events determining the probability of death. Since complex and interacting systems of maintenance and repair comprise the homeodynamic (old term: homeostasis) space of a biological system, aging is considered to be a progressive shrinkage of homeodynamic space mainly due to increased molecular heterogeneity.

Evolutionary Theories

A gene can be expressed at various life-stages. Therefore, natural selection can support lethal and harmful alleles, if their expression occurs after reproduction. Senescence may be the product of such selection. In addition, aging is believed to have evolved because of the increasingly smaller probability of an organism still being alive at older age, due to predation and accidents, both of which may be random and age-invariant. It is thought that strategies that result in a higher reproductive rate at a young age, but shorter overall lifespan, result in a higher lifetime reproductive success and are therefore favoured by natural selection. In essence, aging is, therefore, the result of investing resources in reproduction, rather than maintenance of the body (the "Disposable Soma" theory), in light of the fact that accidents, predation, and disease will eventually kill the organism no matter how much energy is devoted to repair of the body. Various other theories of aging exist, and are not necessarily mutually exclusive.

The geneticist J. B. S. Haldane wondered why the dominant mutation that causes Huntington's disease remained in the population, and why natural selection had not eliminated it. The onset of this neurological disease is (on average) at age 45 and is invariably fatal within 10–20 years. Haldane assumed that, in human prehistory, few survived until age 45. Since few were alive at older ages and their contribution to the next generation was therefore small relative to the large cohorts of younger age groups, the force of selection against such late-acting deleterious mutations was correspondingly small. However, if a mutation affected younger individuals, selection against it would be strong. Therefore, late-acting deleterious mutations could accumulate in populations over evolutionary time through genetic drift. This principle has been demonstrated experimentally. And it is these later-acting deleterious mutations that are believed to cause— even allow—age-related mortality.

Peter Medawar formalised this observation in his mutation accumulation theory of aging. "The force of natural selection weakens with increasing age—even in a theoretically immortal population, provided only that it is exposed to real hazards of mortality. If a genetic disaster... happens late enough in individual life, its consequences may be completely unimportant". The 'real hazards of mortality' are, in typical circumstances, predation, disease, and accidents. So, even an immortal population, whose fertility does not decline with time, will have fewer individuals alive in older age groups. This is called 'extrinsic mortality'. Young cohorts, not depleted in numbers yet by extrinsic mortality, contribute far more to the next generation than the few remaining older cohorts, so the force of selection against late-acting deleterious mutations, which affect only these few older individuals, is very weak. The mutations may not be selected against, therefore, and may spread over evolutionary time into the population.

The major testable prediction made by this model is that species that have high extrinsic mortality in nature will age more quickly and have shorter intrinsic lifespans. This is borne out among mammals, the best-studied in terms of life history. There is a correlation among mammals between body size and lifespan, such that larger species live longer than smaller species in controlled/optimum conditions, but there are notable exceptions. For instance, many bats and rodents are of similar size, yet bats live much longer. For instance, the little brown bat, half the size of a mouse, can live 30 years in the wild. A mouse will live 2–3 years even with optimum conditions. The explanation is that bats have fewer predators, so therefore low extrinsic mortality. Thus, more individuals survive to later ages, so the force of selection against late-acting deleterious mutations is stronger. Fewer late-acting deleterious mutations = slower aging = longer lifespan. Birds are also warm-blooded and are of similar size to many small mammals, yet live often 5–10 times as long. They have fewer predation pressures compared with ground-dwelling mammals. And seabirds, which, in general, have the fewest predators of all birds, live longest.

Also, when examining the body-size vs. lifespan relationship, one will observe that predator mammals tend to have longer lifespans than prey animals in a controlled environment such as a zoo or nature reserve. The explanation for the long lifespans of primates (such as humans, monkeys, and apes) relative to body size is that their intelligence and often sociality helps them avoid becoming prey. Being a predator, being smart, and working together all reduce extrinsic mortality.

Another evolutionary theory of aging was proposed by George C. Williams and involves antagonistic pleiotropy. A single gene may affect multiple traits. Some traits that increase fitness early in life may also have negative effects later in life. But, because many more individuals are alive at young ages than at old ages, even small positive effects early can be strongly selected for, and large negative effects later may be very weakly selected against. Williams suggested the following example: Perhaps a gene codes for calcium deposition in bones, which promotes juvenile survival and will therefore be favored by natural selection; however, this same gene promotes calcium deposition in the arteries, causing negative effects in old age. Therefore, negative effects in old age may reflect the result of natural selection for pleiotropic genes that are beneficial early in life. In this case, fitness is relatively high when Fisher's reproductive value is high and relatively low when Fisher's reproductive value is low.

Gene Regulation

A number of genetic components of aging have been identified using model organisms, ranging from the simple budding yeast *Saccharomyces cerevisiae* to worms such as *Caenorhabditis elegans* and fruit flies (*Drosophila melanogaster*). Study of these organisms has revealed the presence of at least two conserved aging pathways.

One of these pathways involves the gene *Sir2*, a NAD+-dependent histone deacetylase. In yeast, Sir2 is required for genomic silencing at three loci: The yeast mating loci, the telomeres and the ribosomal DNA (rDNA). In some species of yeast, replicative aging may be partially caused by homologous recombination between rDNA repeats; excision of rDNA repeats results in the formation of extrachromosomal rDNA circles (ERCs). These ERCs replicate and preferentially segregate to the mother cell during cell division, and are believed to result in cellular senescence by titrating away (competing for) essential nuclear factors. ERCs have not been observed in other species (nor even all strains of the same yeast species) of yeast (which also display replicative senescence), and ERCs are not believed to contribute to aging in higher organisms such as humans (they have not been shown to accumulate in mammals in a similar manner to yeast). Extrachromosomal circular DNA (eccDNA) has been found in worms, flies, and humans. The origin and role of eccDNA in aging, if any, is unknown.

Despite the lack of a connection between circular DNA and aging in higher organisms, extra copies of Sir2 are capable of extending the lifespan of both worms and flies (though, in flies, this finding has not been replicated by other investigators, and the activator of Sir2 resveratrol does not reproducibly increase lifespan in either species). Whether the Sir2 homologues in higher organisms have any role in lifespan is unclear, but the human SIRT1 protein has been demonstrated to deacetylate p53, Ku70, and the forkhead family of transcription factors. SIRT1 can also regulate acetylates such as CBP/ p300, and has been shown to deacetylate specific histone residues.

RAS1 and RAS2 also affect aging in yeast and have a human homologue. RAS2 overexpression has been shown to extend lifespan in yeast. Other genes regulate aging in yeast by increasing the resistance to oxidative stress. Superoxide dismutase, a protein that protects against the effects of mitochondrial free radicals, can extend yeast lifespan in stationary phase when overexpressed.

In higher organisms, aging is likely to be regulated in part through the insulin/IGF-1 pathway. Mutations that affect insulin-like signaling in worms, flies, and the growth hormone/IGF1 axis in mice are associated with extended lifespan. In yeast, Sir2 activity is regulated by the nicotinamidase PNC1. PNC1 is transcriptionally upregulated under stressful conditions such as caloric restriction, heat shock, and osmotic shock. By converting nicotinamide to niacin, nicotinamide is removed, inhibiting the activity of Sir2. A nicotinamidase found in humans, known as PBEF, may serve a similar function, and a secreted form of PBEF known as visfatin may help to regulate serum insulin levels. It is not known, however, whether these mechanisms also exist in humans, since there are obvious differences in biology between humans and model organisms.

Sir2 activity has been shown to increase under calorie restriction. Due to the lack of available glucose in the cells, more NAD+ is available and can activate Sir2. Resveratrol, a stilbenoid found in the skin of red grapes, was reported to extend the lifespan of yeast, worms, and flies (the lifespan extension in flies and worms have proved irreproducible by independent investigators). It has been shown to activate Sir2 and therefore mimics the effects of calorie restriction, if one accepts that caloric restriction is indeed dependent on Sir2.

Gene expression is imperfectly controlled, and it is possible that random fluctuations in the expression levels of many genes contribute to the aging process as suggested by a study of such genes in yeast. Individual cells, which are genetically identical, none-the-less can have substantially different responses to outside stimuli, and markedly different lifespans, indicating the epigenetic factors play an important role in gene expression and aging as well as genetic factors.

According to the GenAge database of aging-related genes there are over 700 genes associated with aging in model organisms: 555 in the soil roundworm (*Caenorhabditis elegans*), 87 in the bakers' yeast (*Saccharomyces cerevisiae*), 75 in the fruit fly (*Drosophila melanogaster*) and 68 in the mouse (*Mus musculus*).

Cellular Senescence

As noted above, senescence is not universal, and senescence is not observed in single-celled organisms that reproduce through the process of cellular mitosis. Moreover, cellular senescence is not observed in several organisms, including perennial plants, sponges, corals, and lobsters. In those species where cellular senescence is observed, cells eventually become post-mitotic when they can no longer replicate themselves through the process of cellular mitosis; i.e., cells experience *replicative senescence*. How and why some cells become post-mitotic in some species has been the subject of much research and speculation, but (as noted above) it is widely believed that cellular senescence evolved as a way to prevent the onset and spread of cancer. Somatic cells that have divided many times will have accumulated DNA mutations and would therefore be in danger of becoming cancerous if cell division continued.

Lately, the role of telomeres in cellular senescence has aroused general interest, especially with a view to the possible genetically adverse effects of cloning. The successive shortening of the chromosomal telomeres with each cell cycle is also believed to limit the number of divisions of the cell, thus contributing to aging. There have, on the other hand, also been reports that cloning could alter the shortening of telomeres. Some cells do not age and are, therefore, described as being "biologically immortal". It is theorized by some that when it is discovered exactly what allows these cells, whether it be the result of telomere lengthening or not, to divide without limit that it will be possible to genetically alter other cells to have the same capability. It is further theorized that it will eventually be possible to genetically engineer all cells in the human body to have this capability by employing gene therapy and, therefore, stop or reverse aging, effectively making the entire organism potentially immortal.

Cancer cells are usually immortal. In about 85% of tumors, this evasion of cellular senescence is the result of up-activation of their telomerase genes. This simple observation suggests that reactivation of telomerase in healthy individuals could greatly increase their cancer risk. Whether cell senescence plays any role in organismal aging is at present unknown, and is an active area of investigation.

Mouse mutants lacking telomerase do not immediately show accelerated aging.

Chemical Damage

One of the earliest aging theories was the *Rate of Living Hypothesis* described by Raymond Pearl in 1928(based on earlier work by Max Rubner), which states that fast basal metabolic rate corresponds to short maximum life span.

While there may be some validity to the idea that for various types of specific damage detailed below that are by-products of metabolism, all other things being equal, a fast metabolism may reduce lifespan, in general this theory does not adequately explain the differences in lifespan either within, or between, species.

Calorically-restricted animals process as much, or more, calories per gram of body mass, as their *ad libitum* fed counterparts, yet exhibit substantially longer lifespans.. Similarly, metabolic rate is a poor predictor of lifespan for birds, bats and other species that, it is presumed, have reduced mortality from predation, and therefore have evolved long lifespans even in the presence of very high metabolic rates. More recently, it was shown that, when modern statistical methods for correcting for the effects of body size and phylogeny are employed, metabolic rate does not correlate with longevity in mammals or birds.

With respect to specific types of chemical damage caused by metabolism, it is suggested that damage to long-lived biopolymers, such as structural proteins or DNA, caused by ubiquitous chemical agents in the body such as oxygen and sugars, are in part responsible for aging. The damage can include breakage of biopolymer chains, cross-linking of biopolymers, or chemical attachment of unnatural substituents (haptens) to biopolymers.

Under normal aerobic conditions, approximately 4% of the oxygen metabolized by mitochondria is converted to superoxide ion, which can subsequently be converted to hydrogen peroxide, hydroxyl radical and eventually other reactive species including other peroxides and singlet oxygen, which can, in turn, generate free radicals capable of damaging structural proteins and DNA. Certain metal ions found in the body, such as copper and iron, may participate in the process. (In Wilson's disease, a hereditary defect that causes the body to retain copper, some of the symptoms resemble accelerated senescence.) These processes are termed *oxidative damage* and are linked to the benefits of nutritionally derived polyphenol antioxidants.

Sugars such as glucose and fructose can react with certain amino acids such as lysine and arginine and certain DNA bases such as guanine to produce sugar adducts, in a process called *glycation*. These adducts can further rearrange to form reactive species, which can then cross-link the structural proteins or DNA to similar biopolymers or other biomolecules such as non-structural proteins. People with diabetes, who have elevated blood sugar, develop senescence-associated disorders much earlier than the general population, but can delay such disorders by rigorous control of their blood sugar levels. There is evidence that sugar damage is linked to oxidant damage in a process termed *glycoxidation*.

Free radicals can damage proteins, lipids or DNA. Glycation mainly damages proteins. Damaged proteins and lipids accumulate in lysosomes as lipofuscin. Chemical damage to structural proteins can lead to loss of function; for example, damage to collagen of blood vessel walls can lead to vessel-wall stiffness and, thus, hypertension, and vessel wall thickening and reactive tissue formation (atherosclerosis); similar processes in the kidney can lead to renal failure. Damage to enzymes reduces cellular functionality. Lipid peroxidation of the inner mitochondrial membrane reduces the electric potential and the ability to generate energy. It is probably no accident that nearly all of the so-called "accelerated aging diseases" are due to defective DNA repair enzymes.

It is believed that the impact of alcohol on aging can be partly explained by alcohol's activation of the HPA axis, which stimulates glucocorticoid secretion, long-term exposure to which produces symptoms of aging.

Reliability Theory

Reliability theory suggests that biological systems start their adult life with a high load of initial damage. Reliability theory is a general theory about systems failure. It allows researchers to predict the age-related failure kinetics for a system of given architecture (reliability structure) and given reliability of its components. Reliability theory predicts that even those systems that composed entirely of non-aging elements (with a constant failure rate) will nevertheless deteriorate (fail more often) with age, if these systems are redundant in irreplaceable elements. Aging, therefore, is a direct consequence of systems.

Reliability theory also predicts the late-life mortality deceleration with subsequent leveling-off, as well as the late-life mortality plateaus, as an inevitable consequence of redundancy exhaustion at extreme old ages.

The theory explains why mortality rates increase exponentially with age (the Gompertz law) in many species, by taking into account the initial flaws (defects) in newly formed systems. It also explains why organisms "prefer" to die according to the Gompertz law, while technical devices usually fail according to the Weibull (power) law. Reliability theory allows to specify conditions when organisms die according to the Weibull distribution: Organisms should be relatively free of initial flaws and defects.

The theory makes it possible to find a general failure law applicable to all adult and extreme old ages, where the Gompertz and the Weibull laws are just special cases of this more general failure law. The theory explains why relative differences in mortality rates of compared populations (within a given species) vanish with age (compensation law of mortality), and mortality convergence is observed due to the exhaustion of initial differences in redundancy levels.

Miscellaneous

Recently, a kind of early senescence has been alleged to be a possible unintended outcome of early cloning experiments. The issue was raised in the case of Dolly the sheep, following her death from a contagious lung disease. The claim that Dolly's early death involved premature senescence has been vigorously contested, and Dolly's creator, Dr. Ian Wilmut has expressed the view that her illness and death were probably unrelated to the fact that she was a clone.

A set of rare hereditary (genetic) disorders, each called progeria, has been known for some time. Sufferers exhibit symptoms resembling accelerated aging, including wrinkled skin. The cause of Hutchinson–Gilford progeria syndrome was reported in the journal *Nature* in May 2003. This report suggests that DNA damage, not oxidative stress, is the cause of this form of accelerated aging.

Transcription

Transcription is the process of creating a complementary RNA copy of a sequence of DNA. Both RNA and DNA are nucleic acids, which use base pairs of nucleotides as a complementary language that can be converted back and forth from DNA to RNA by the action of the correct enzymes. During transcription, a DNA sequence is read by RNA polymerase, which produces a complementary, antiparallel RNA strand. As opposed to DNA replication, transcription results in an RNA complement that includes uracil (U) in all instances where thymine (T) would have occurred in a DNA complement.

Transcription can be explained easily in 4 or 5 simple steps, each moving like a wave along the DNA.

1. RNA polymerase unwinds/"unzips" the DNA by breaking the hydrogen bonds between complimentary nucleotides.

2. RNA nucleotides are paired with complementary DNA bases.

3. RNA sugar-phosphate backbone forms with assistance from RNA polymerase.

4. Hydrogen bonds of the untwisted RNA+DNA helix break, freeing the newly synthesized RNA strand.

5. If the cell has a nucleus, the RNA is further processed and then moves through the small nuclear pores to the cytoplasm.

Transcription is the first step leading to gene expression. The stretch of DNA transcribed into an RNA molecule is called a *transcription unit* and encodes at least one gene. If the gene transcribed encodes a protein, the result of transcription is messenger RNA (mRNA), which will then be used to create that protein via the process of translation. Alternatively, the transcribed gene may encode for either ribosomal RNA (rRNA) or transfer RNA (tRNA), other components of the protein-assembly process, or other ribozymes.

A DNA transcription unit encoding for a protein contains not only the sequence that will eventually be directly translated into the protein (the *coding sequence*) but also *regulatory sequences* that direct and regulate the synthesis of that protein. The regulatory sequence before (upstream from) the coding sequence is called the five prime untranslated region (5'UTR), and the sequence following (downstream from) the coding sequence is called the three prime untranslated region (3'UTR).

Transcription has some proofreading mechanisms, but they are fewer and less effective than the controls for copying DNA; therefore, transcription has a lower copying fidelity than DNA replication.

As in DNA replication, DNA is read from 3' → 5' during transcription. Meanwhile, the complementary RNA is created from the 5' '! 3' direction. This means its 5' end is created first in base pairing. Although DNA is arranged as two antiparallel strands in a double helix, only one of the two DNA strands, called the template strand, is used for transcription. This is because RNA is only single-stranded, as opposed to double-stranded DNA. The other DNA strand is called the coding strand, because its sequence is the same as the newly created RNA transcript (except for the substitution of uracil for thymine). The use of only the 3' '! 5' strand eliminates the need for the Okazaki fragments seen in DNA replication.

Transcription is divided into 5 stages: *pre-initiation, initiation, promoter clearance, elongation* and *termination*.

Major Steps

Pre-initiation

In eukaryotes, RNA polymerase, and therefore the initiation of transcription, requires the presence of a core promoter sequence in the DNA. Promoters are regions of DNA that promote transcription and, in eukaryotes, are found at -30, -75, and -90 base pairs upstream from the start site of transcription. Core promoters are sequences within the promoter that are essential for transcription initiation. RNA polymerase is able to bind to core promoters in the presence of various specific transcription factors.

The most common type of core promoter in eukaryotes is a short DNA sequence known as a TATA box, found 25-30 base pairs upstream from the start site of transcription. The TATA box, as a core promoter, is the binding site for a transcription factor known as TATA-binding protein (TBP), which is itself a subunit of another transcription factor, called Transcription Factor II D (TFIID). After TFIID binds to the TATA box via the TBP, five more transcription factors and RNA polymerase combine around the TATA box in a series of stages to form a preinitiation complex. One transcription factor, DNA helicase, has helicase activity and so is involved in the separating of opposing strands of double-stranded DNA to provide access to a single-stranded DNA template. However, only a low, or basal, rate of transcription is driven by the preinitiation complex alone. Other proteins known as activators and repressors, along with any associated coactivators or corepressors, are responsible for modulating transcription rate.

Thus, preinitiation complex contains: 1. Core Promoter Sequence 2. Transcription Factors 3. DNA Helicase 4. RNA Polymerase 5. Activators and Repressors The transcription preinitiation in archaea is, in essence, homologous to that of eukaryotes, but is much less complex. The archaeal preinitiation complex assembles at a TATA-box binding site; however, in archaea, this complex is composed of only RNA polymerase II, TBP, and TFB (the archaeal homologue of eukaryotic transcription factor II B (TFIIB)).

Initiation

In bacteria, transcription begins with the binding of RNA polymerase to the promoter in DNA. RNA polymerase is a core enzyme consisting of five subunits: 2 α subunits, 1 β subunit, 1 β' subunit, and 1 ω subunit. At the start of initiation, the core enzyme is associated with a sigma factor that aids in finding the appropriate -35 and -10 base pairs downstream of promoter sequences.

Transcription initiation is more complex in eukaryotes. Eukaryotic RNA polymerase does not directly recognize the core promoter sequences. Instead, a collection of proteins called transcription factors mediate the binding of RNA polymerase and the initiation of transcription. Only after certain transcription factors are attached to the promoter does the RNA polymerase bind to it. The completed assembly of transcription factors and RNA polymerase bind to the promoter, forming a transcription initiation complex. Transcription in the archaea domain is similar to transcription in eukaryotes.

Promoter Clearance

After the first bond is synthesized, the RNA polymerase must clear the promoter. During this time there is a tendency to release the RNA transcript and produce truncated transcripts. This is called *abortive initiation* and is common for both eukaryotes and prokaryotes. Abortive initiation continues to occur until the σ factor rearranges, resulting in the transcription elongation complex (which gives a 35 bp moving footprint). The σ factor is released before 80 nucleotides of mRNA are synthesized. Once the transcript reaches approximately 23 nucleotides, it no longer slips and elongation can occur. This, like most of the remainder of transcription, is an energy-dependent process, consuming adenosine triphosphate (ATP).

Promoter clearance coincides with phosphorylation of serine 5 on the carboxy terminal domain of RNA Pol in eukaryotes, which is phosphorylated by TFIIH.

Elongation

One strand of the DNA, the *template strand* (or noncoding strand), is used as a template for RNA synthesis. As transcription proceeds,

RNA polymerase traverses the template strand and uses base pairing complementarity with the DNA template to create an RNA copy. Although RNA polymerase traverses the template strand from 3' '! 5', the coding (non-template) strand and newly-formed RNA can also be used as reference points, so transcription can be described as occurring 5' '! 3'. This produces an RNA molecule from 5' '! 3', an exact copy of the coding strand (except that thymines are replaced with uracils, and the nucleotides are composed of a ribose (5-carbon) sugar where DNA has deoxyribose (one less oxygen atom) in its sugar-phosphate backbone).

Unlike DNA replication, mRNA transcription can involve multiple RNA polymerases on a single DNA template and multiple rounds of transcription (amplification of particular mRNA), so many mRNA molecules can be rapidly produced from a single copy of a gene.

Elongation also involves a proofreading mechanism that can replace incorrectly incorporated bases. In eukaryotes, this may correspond with short pauses during transcription that allow appropriate RNA editing factors to bind. These pauses may be intrinsic to the RNA polymerase or due to chromatin structure.

Termination

Bacteria use two different strategies for transcription termination. In Rho-independent transcription termination, RNA transcription stops when the newly synthesized RNA molecule forms a G-C-rich hairpin loop followed by a run of Us. When the hairpin forms, the mechanical stress breaks the weak rU-dA bonds, now filling the DNA-RNA hybrid. This pulls the poly-U transcript out of the active site of the RNA polymerase, in effect, terminating transcription. In the "Rho-dependent" type of termination, a protein factor called "Rho" destabilizes the interaction between the template and the mRNA, thus releasing the newly synthesized mRNA from the elongation complex.

Transcription termination in eukaryotes is less understood but involves cleavage of the new transcript followed by template-independent addition of As at its new 3' end, in a process called polyadenylation.

Measuring and Detecting Transcription

Transcription can be measured and detected in a variety of ways:
  • Nuclear Run-on assay: measures the relative abundance of newly formed transcripts

- RNase protection assay and ChIP-Chip of RNAP: detect active transcription sites
- RT-PCR: measures the absolute abundance of total or nuclear RNA levels, which may however differ from transcription rates
- DNA microarrays: measures the relative abundance of the global total or nuclear RNA levels; however, these may differ from transcription rates
- In situ hybridization: detects the presence of a transcript
- MS2 tagging: by incorporating RNA stem loops, such as MS2, into a gene, these become incorporated into newly synthesized RNA. The stem loops can then be detected using a fusion of GFP and the MS2 coat protein, which has a high affinity, sequence-specific interaction with the MS2 stem loops. The recruitment of GFP to the site of transcription is visualised as a single fluorescent spot. This remarkable new approach has revealed that transcription occurs in discontinuous bursts, or pulses. With the notable exception of in situ techniques, most other methods provide cell population averages, and are not capable of detecting this fundamental property of genes.
- Northern blot: the traditional method, and until the advent of RNA-Seq, the most quantitative
- RNA-Seq: applies next-generation sequencing techniques to sequence whole transcriptomes, which allows the measurement of relative abundance of RNA, as well as the detection of additional variations such as fusion genes, post-translational edits and novel splice sites.

Transcription Factories

Active transcription units are clustered in the nucleus, in discrete sites called transcription factories or euchromatin. Such sites can be visualized by allowing engaged polymerases to extend their transcripts in tagged precursors (Br-UTP or Br-U) and immuno-labelling the tagged nascent RNA. Transcription factories can also be localized using fluorescence in situ hybridization or marked by antibodies directed against polymerases. There are ~10,000 factories in the nucleoplasm of a HeLa cell, among which are ~8,000 polymerase II factories and ~2,000 polymerase III factories. Each polymerase II factory contains ~8 polymerases. As most active transcription units are associated with only one polymerase, each factory usually contains

~8 different transcription units. These units might be associated through promoters and/or enhancers, with loops forming a 'cloud' around the factor.

History

A molecule that allows the genetic material to be realized as a protein was first hypothesized by François Jacob and Jacques Monod. RNA synthesis by RNA polymerase was established *in vitro* by several laboratories by 1965; however, the RNA synthesized by these enzymes had properties that suggested the existence of an additional factor needed to terminate transcription correctly.

In 1972, Walter Fiers became the first person to actually prove the existence of the terminating enzyme.

Roger D. Kornberg won the 2006 Nobel Prize in Chemistry "for his studies of the molecular basis of eukaryotic transcription".

Reverse Transcription

Some viruses (such as HIV, the cause of AIDS), have the ability to transcribe RNA into DNA. HIV has an RNA genome that is duplicated into DNA. The resulting DNA can be merged with the DNA genome of the host cell. The main enzyme responsible for synthesis of DNA from an RNA template is called reverse transcriptase. In the case of HIV, reverse transcriptase is responsible for synthesizing a complementary DNA strand (cDNA) to the viral RNA genome. An associated enzyme, ribonuclease H, digests the RNA strand, and reverse transcriptase synthesises a complementary strand of DNA to form a double helix DNA structure. This cDNA is integrated into the host cell's genome via another enzyme (integrase) causing the host cell to generate viral proteins that reassemble into new viral particles. In HIV, subsequent to this, the host cell undergoes programmed cell death, apoptosis of T cells. However, in other retroviruses, the host cell remains intact as the virus buds out of the cell.

Some eukaryotic cells contain an enzyme with reverse transcription activity called telomerase. Telomerase is a reverse transcriptase that lengthens the ends of linear chromosomes. Telomerase carries an RNA template from which it synthesizes DNA repeating sequence, or "junk" DNA. This repeated sequence of DNA is important because, every time a linear chromosome is duplicated, it is shortened in length. With "junk" DNA at the ends of chromosomes, the shortening eliminates some of the non-essential, repeated sequence rather than the protein-encoding DNA sequence farther away from the chromosome end. Telomerase is often activated in cancer cells to enable cancer cells to duplicate their genomes indefinitely without losing important protein-coding DNA sequence. Activation of telomerase could be part of the process that allows cancer cells to become *immortal*. However, the true *in vivo* significance of telomerase has still not been empirically proven.

Ribonucleic Acid

Ribonucleic acid (RNA) is one of the three major macromolecules (along with DNA and proteins) that are essential for all known forms of life.

Like DNA, RNA is made up of a long chain of components called nucleotides. Each nucleotide consists of a nucleobase (sometimes called a nitrogenous base), a ribose sugar, and a phosphate group. The sequence of nucleotides allows RNA to encode genetic information. For example, some viruses use RNA instead of DNA as their genetic material, and all organisms use messenger RNA (mRNA) to carry the genetic information that directs the synthesis of proteins.

Like proteins, some RNA molecules play an active role in cells by catalyzing biological reactions, controlling gene expression, or sensing and communicating responses to cellular signals. One of these active processes is protein synthesis, a universal function whereby mRNA molecules direct the assembly of proteins on ribosomes. This process uses transfer RNA (tRNA) molecules to deliver amino acids to the ribosome, where ribosomal RNA (rRNA) links amino acids together to form proteins.

The chemical structure of RNA is very similar to that of DNA, with two differences—(a) RNA contains the sugar ribose while DNA contains the slightly different sugar deoxyribose (a type of ribose that lacks one oxygen atom), and (b) RNA has the nucleobase uracil while DNA contains thymine (uracil and thymine have similar base-pairing properties). Unlike DNA, most RNA molecules are single-stranded. Single-stranded RNA molecules adopt very complex three-dimensional structures, since they are not restricted to the repetitive double-helical form of double-stranded DNA. RNA is made within living cells by RNA polymerases, enzymes that act to copy a DNA or RNA template into a new RNA strand through processes known as transcription or RNA replication, respectively.

Comparison with DNA

RNA and DNA are both nucleic acids, but differ in three main ways. First, unlike DNA, which is, in general, double-stranded, RNA is a single-stranded molecule in many of its biological roles and has a much shorter chain of nucleotides. Second, while DNA contains *deoxyribose*, RNA contains *ribose* (in deoxyribose there is no hydroxyl group attached to the pentose ring in the 2' position). These hydroxyl groups make RNA less stable than DNA because it is more prone to hydrolysis. Third, the complementary base to adenine is not thymine, as it is in DNA, but rather uracil, which is an unmethylated form of thymine.

Like DNA, most biologically active RNAs, including mRNA, tRNA, rRNA, snRNAs, and other non-coding RNAs, contain self-complementary sequences that allow parts of the RNA to fold and pair with itself to form double helices. Structural analysis of these RNAs has revealed that they are highly structured. Unlike DNA, their structures do not consist of long double helices but rather collections of short helices packed together into structures akin to proteins. In this fashion, RNAs can achieve chemical catalysis, like enzymes. For instance, determination of the structure of the ribosome—an enzyme that catalyzes peptide bond formation—revealed that its active site is composed entirely of RNA.

Structure

Each nucleotide in RNA contains a ribose sugar, with carbons numbered 1' through 5'. A base is attached to the 1' position, in general, adenine (A), cytosine (C), guanine (G), or uracil (U). Adenine and guanine are purines, cytosine, and uracil are pyrimidines. A phosphate group is attached to the 3' position of one ribose and the 5' position of the next. The phosphate groups have a negative charge each at physiological pH, making RNA a charged molecule (polyanion). The bases may form hydrogen bonds between cytosine and guanine, between adenine and uracil and between guanine and uracil. However, other interactions are possible, such as a group of adenine bases binding to each other in a bulge, or the GNRA tetraloop that has a guanine–adenine base-pair.

An important structural feature of RNA that distinguishes it from DNA is the presence of a hydroxyl group at the 2' position of the ribose sugar. The presence of this functional group causes the helix to adopt the A-form geometry rather than the B-form most commonly observed in DNA. This results in a very deep and narrow major groove and a shallow and wide minor groove. A second consequence of the presence of the 2'-hydroxyl group is that in conformationally flexible regions of an RNA molecule (that is, not involved in formation of a double helix), it can chemically attack the adjacent phosphodiester bond to cleave the backbone.

RNA is transcribed with only four bases (adenine, cytosine, guanine and uracil), but these bases and attached sugars can be modified in numerous ways as the RNAs mature. Pseudouridine (Ø), in which the linkage between uracil and ribose is changed from a C–N bond to a C–C bond, and ribothymidine (T) are found in various places (the most notable ones being in the TØC loop of tRNA). Another notable modified base is hypoxanthine, a deaminated adenine base whose nucleoside is called inosine (I). Inosine plays a key role in the wobble hypothesis of the genetic code.

There are nearly 100 other naturally occurring modified nucleosides, of which pseudouridine and nucleosides with 2'-O-methylribose are the most common. The specific roles of many of these modifications in RNA are not fully understood. However, it is notable that, in ribosomal RNA, many of the post-transcriptional modifications occur in highly functional regions, such as the peptidyl transferase center and the subunit interface, implying that they are important for normal function.

The functional form of single stranded RNA molecules, just like proteins, frequently requires a specific tertiary structure.

The scaffold for this structure is provided by secondary structural elements that are hydrogen bonds within the molecule. This leads to several recognizable "domains" of secondary structure like hairpin loops, bulges, and internal loops. Since RNA is charged, metal ions such as Mg^{2+} are needed to stabilise many secondary and tertiary structures.

Synthesis

Synthesis of RNA is usually catalyzed by an enzyme—RNA polymerase—using DNA as a template, a process known as transcription. Initiation of transcription begins with the binding of the enzyme to a promoter sequence in the DNA (usually found "upstream" of a gene). The DNA double helix is unwound by the helicase activity of the enzyme. The enzyme then progresses along the template strand in the 3' to 5' direction, synthesizing a complementary RNA molecule with elongation occurring in the 5' to 3' direction. The DNA sequence also dictates where termination of RNA synthesis will occur.

RNAs are often modified by enzymes after transcription. For example, a poly(A) tail and a 5' cap are added to eukaryotic pre-mRNA and introns are removed by the spliceosome.

There are also a number of RNA-dependent RNA polymerases that use RNA as their template for synthesis of a new strand of RNA. For instance, a number of RNA viruses (such as poliovirus) use this type of enzyme to replicate their genetic material. Also, RNA-dependent RNA polymerase is part of the RNA interference pathway in many organisms.

Types of RNA

Messenger RNA (mRNA) is the RNA that carries information from DNA to the ribosome, the sites of protein synthesis (translation) in the cell. The coding sequence of the mRNA determines the amino acid sequence in the protein that is produced. Many RNAs do not code for protein however (about 97% of the transcriptional output is non-protein-coding in eukaryotes).

These so-called non-coding RNAs ("ncRNA") can be encoded by their own genes (RNA genes), but can also derive from mRNA introns. The most prominent examples of non-coding RNAs are transfer RNA (tRNA) and ribosomal RNA (rRNA), both of which are involved in the process of translation.

There are also non-coding RNAs involved in gene regulation, RNA processing and other roles. Certain RNAs are able to catalyse chemical reactions such as cutting and ligating other RNA molecules, and the catalysis of peptide bond formation in the ribosome; these are known as ribozymes.

In Translation

Messenger RNA (mRNA) carries information about a protein sequence to the ribosomes, the protein synthesis factories in the cell. It is coded so that every three nucleotides (a codon) correspond to one amino acid. In eukaryotic cells, once precursor mRNA (pre-mRNA) has been transcribed from DNA, it is processed to mature mRNA. This removes its introns—non-coding sections of the pre-mRNA. The mRNA is then exported from the nucleus to the cytoplasm, where it is bound to ribosomes and translated into its corresponding protein form with the help of tRNA. In prokaryotic cells, which do not have nucleus and cytoplasm compartments, mRNA can bind to ribosomes while it is being transcribed from DNA. After a certain amount of time the message degrades into its component nucleotides with the assistance of ribonucleases.

Transfer RNA (tRNA) is a small RNA chain of about 80 nucleotides that transfers a specific amino acid to a growing polypeptide chain at the ribosomal site of protein synthesis during translation. It has sites for amino acid attachment and an anticodon region for codon recognition that binds to a specific sequence on the messenger RNA chain through hydrogen bonding.

Ribosomal RNA (rRNA) is the catalytic component of the ribosomes. Eukaryotic ribosomes contain four different rRNA molecules: 18S, 5.8S, 28S and 5S rRNA. Three of the rRNA molecules are synthesized in the nucleolus, and one is synthesized elsewhere. In the cytoplasm, ribosomal RNA and protein combine to form a nucleoprotein called a ribosome. The ribosome binds mRNA and carries out protein synthesis. Several ribosomes may be attached to a single mRNA at any time. rRNA is extremely abundant and makes up 80% of the 10 mg/ml RNA found in a typical eukaryotic cytoplasm.

Transfer-messenger RNA (tmRNA) is found in many bacteria and plastids. It tags proteins encoded by mRNAs that lack stop codons for degradation and prevents the ribosome from stalling.

Regulatory RNAs

Several types of RNA can downregulate gene expression by being complementary to a part of an mRNA or a gene's DNA. MicroRNAs (miRNA; 21-22 nt) are found in eukaryotes and act through RNA interference (RNAi), where an effector complex of miRNA and enzymes can break down mRNA to which the miRNA is complementary, block the mRNA from being translated, or accelerate its degradation. While small interfering RNAs (siRNA; 20-25 nt) are often produced by breakdown of viral RNA, there are also endogenous sources of siRNAs.

siRNAs act through RNA interference in a fashion similar to miRNAs. Some miRNAs and siRNAs can cause genes they target to be methylated, thereby decreasing or increasing transcription of those genes. Animals have Piwi-interacting RNAs (piRNA; 29-30 nt) which are active in germline cells and are thought to be a defense against transposons and play a role in gametogenesis.

Many prokaryotes have CRISPR RNAs, a regulatory system similar to RNA interference. Antisense RNAs are widespread; most downregulate a gene, but a few are activators of transcription. One way antisense RNA can act is by binding to an mRNA, forming double-stranded RNA that is enzymatically degraded. There are many long noncoding RNAs that regulate genes in eukaryotes, one such RNA is Xist, which coats one X chromosome in female mammals and inactivates it.

An mRNA may contain regulatory elements itself, such as riboswitches, in the 5' untranslated region or 3' untranslated region; these cis-regulatory elements regulate the activity of that mRNA. The untranslated regions can also contain elements that regulate other genes.

In RNA Processing

Many RNAs are involved in modifying other RNAs. Introns are spliced out of pre-mRNA by spliceosomes, which contain several small nuclear RNAs (snRNA), or the introns can be ribozymes that are spliced by themselves. RNA can also be altered by having its nucleotides modified to other nucleotides than A, C, G and U. In eukaryotes, modifications of RNA nucleotides are generally directed by small nucleolar RNAs (snoRNA; 60-300 nt), found in the nucleolus and cajal bodies. snoRNAs associate with enzymes and guide them to a spot on an RNA by basepairing to that RNA. These enzymes then perform the nucleotide modification. rRNAs and tRNAs are extensively modified, but snRNAs and mRNAs can also be the target of base modification.

RNA Genomes

Like DNA, RNA can carry genetic information. RNA viruses have genomes composed of RNA, and a variety of proteins encoded by that genome. The viral genome is replicated by some of those proteins, while other proteins protect the genome as the virus particle moves to a new host cell. Viroids are another group of pathogens, but they consist only of RNA, do not encode any protein and are replicated by a host plant cell's polymerase.

In Reverse Transcription

Reverse transcribing viruses replicate their genomes by reverse transcribing DNA copies from their RNA; these DNA copies are then transcribed to new RNA. Retrotransposons also spread by copying DNA and RNA from one another, and telomerase contains an RNA that is used as template for building the ends of eukaryotic chromosomes.

Double-stranded RNA

Double-stranded RNA (dsRNA) is RNA with two complementary strands, similar to the DNA found in all cells. dsRNA forms the genetic material of some viruses (double-stranded RNA viruses). Double-stranded RNA such as viral RNA or siRNA can trigger RNA interference in eukaryotes, as well as interferon response in vertebrates.

Key Discoveries in RNA Biology

Research on RNA has led to many important biological discoveries and numerous Nobel Prizes. Nucleic acids were discovered in 1868 by Friedrich Miescher, who called the material 'nuclein' since it was found in the nucleus. It was later discovered that prokaryotic cells, which do not have a nucleus, also contain nucleic acids. The role of RNA in protein synthesis was suspected already in 1939. Severo Ochoa won the 1959 Nobel Prize in Medicine (shared with Arthur Kornberg) after he discovered an enzyme that can synthesize RNA in the laboratory. Ironically, the enzyme discovered by Ochoa (polynucleotide phosphorylase) was later shown to be responsible for RNA degradation, not RNA synthesis.

The sequence of the 77 nucleotides of a yeast tRNA was found by Robert W. Holley in 1965, winning Holley the 1968 Nobel Prize in Medicine (shared with Har Gobind Khorana and Marshall Nirenberg). In 1967, Carl Woese hypothesized that RNA might be catalytic and suggested that the earliest forms of life (self-replicating molecules) could have relied on RNA both to carry genetic information and to catalyze biochemical reactions—an RNA world.

During the early 1970s, retroviruses and reverse transcriptase were discovered, showing for the first time that enzymes could copy RNA into DNA (the opposite of the usual route for transmission of genetic information). For this work, David Baltimore, Renato Dulbecco and Howard Temin were awarded a Nobel Prize in 1975. In 1976, Walter Fiers and his team determined the first complete nucleotide sequence of an RNA virus genome, that of bacteriophage MS2.

In 1977, introns and RNA splicing were discovered in both mammalian viruses and in cellular genes, resulting in a 1993 Nobel to Philip Sharp and Richard Roberts. Catalytic RNA molecules (ribozymes) were discovered in the early 1980s, leading to a 1989 Nobel award to Thomas Cech and Sidney Altman. In 1990 it was found in petunia that introduced genes can silence similar genes of the plant's own, now known to be a result of RNA interference.

At about the same time, 22 nt long RNAs, now called microRNAs, were found to have a role in the development of *C. elegans*. Studies on RNA interference gleaned a Nobel Prize for Andrew Fire and Craig Mello in 2006, and another Nobel was awarded for studies on transcription of RNA to Roger Kornberg in the same year. The discovery of gene regulatory RNAs has led to attempts to develop drugs made of RNA, such as siRNA, to silence genes.

Transcription and Processing of RNA

Gene Expression: Transcription

The majority of genes are expressed as the proteins they encode. The process occurs in two steps:

- Transcription = DNA → RNA
- Translation = RNA → protein.

Taken together, they make up the "central dogma" of biology: DNA → RNA → protein.

Gene Transcription: DNA → RNA

DNA serves as the template for the synthesis of RNA much as it does for its own replication.

The Steps

- Some 50 different protein transcription factors bind to promoter sites, usually on the 5' side of the gene to be transcribed.
- An enzyme, an RNA polymerase, binds to the complex of transcription factors.
- Working together, they open the DNA double helix.
- The RNA polymerase proceeds to "read" one strand moving in its 3' → 5' direction.
- In eukaryotes, this requires — at least for protein-encoding genes — that the nucleosomes in front of the advancing RNA polymerase (RNAP II) be removed. A complex of proteins is responsible for this. The same complex replaces the nucleosomes after the DNA has been transcribed and RNAP II has moved on.
- As the RNA polymerase travels along the DNA strand, it assembles ribonucleotides (supplied as triphosphates, e.g., ATP) into a strand of RNA.

- Each ribonucleotide is inserted into the growing RNA strand following the rules of base pairing. Thus for each C encountered on the DNA strand, a G is inserted in the RNA; for each G, a C; and for each T, an A. However, each A on the DNA guides the insertion of the pyrimidine uracil (U, from uridine triphosphate, UTP). There is no T in RNA.

Quality control: Occasionally RNA polymerase will select and insert an incorrect, mismatched, ribonucleotide. When this occurs in bacteria (and perhaps in all organisms), the enzyme backs up, removes the incorrect nucleotide (and the one preceding it) and tries again. (Described by Zenkin *et al.*, in the 28 July 2006 issue of Science.)

- Synthesis of the RNA proceeds in the 52 '! 32 direction.
- As each nucleoside triphosphate is brought in to add to the 32 end of the growing strand, the two terminal phosphates are removed.
- When transcription is complete, the transcript is released from the polymerase and, shortly thereafter, the polymerase is released from the DNA.

Note that at any place in a DNA molecule, either strand may be serving as the template; that is, some genes "run" one way, some the other (and in a few remarkable cases, the same segment of double helix contains genetic information on both strands!). In all cases, however, RNA polymerase transcribes the DNA strand in its 32 '! 52 direction.

A report in the 4 January 2001 issue of Nature shows that RNA polymerase actually tracks around the double helix of DNA. In vitro, at least, when RNA polymerase is immobilized, it spins the DNA molecule around and around as it moves along the molecule. Whether it is the polymerase or the DNA that does the spinning in vivo remains to be determined.

Types of RNA

Sedimentation pattern produced by high-speed centrifugation of RNA extracted from the precursors of rabbit red blood cells. The discrete bands represent particular classes of RNA.

The transfer RNAs band at about 4S. The ribosomal RNAs of eukaryotes sediment at 5S, 5.8S, 18S, and 28S. (The larger the sedimentation unit, S, the larger the molecule — but not proportionally.) The RNA forming the band at 9S is the messenger RNA for the synthesis of hemoglobin, the major protein synthesized in these cells. In most types of cells, the messenger RNAs are extremely heterogenous, with small amounts distributed from 6S to 25S.

Several types of RNA are synthesized in the nucleus of eukaryotic cells. Of particular interest are:

- messenger RNA (mRNA). This will later be translated into a polypeptide.
- ribosomal RNA (rRNA). This will be used in the building of ribosomes: machinery for synthesizing proteins by translating mRNA.
- transfer RNA (tRNA). RNA molecules that carry amino acids to the growing polypeptide.
- small nuclear RNA (snRNA). DNA transcription of the genes for mRNA, rRNA, and tRNA produces large precursor molecules ("primary transcripts") that must be processed within the nucleus to produce the functional molecules for export to the cytosol. Some of these processing steps are mediated by snRNAs.
- small nucleolar RNA (snoRNA). These RNAs within the nucleolus have several functions (described below).
- microRNA (miRNA). These are tiny (~22 nucleotides) RNA molecules that appear to regulate the expression of messenger RNA (mRNA) molecules. [Discussion]
- XIST RNA. This inactivates one of the two X chromosomes in female vertebrates. [Discussion].

Messenger RNA (mRNA)

Messenger RNA comes in a wide range of sizes reflecting the size of the polypeptide it encodes. Most cells produce small amounts of thousands of different mRNA molecules, each to be translated into a peptide needed by the cell.

Many mRNAs are common to most cells, encoding "housekeeping" proteins needed by all cells (e.g. the enzymes of glycolysis). Other mRNAs are specific for only certain types of cells. These encode proteins needed for the function of that particular cell (e.g., the mRNA for hemoglobin in the precursors of red blood cells).

Ribosomal RNA (rRNA)

There are 4 kinds. In eukaryotes, these are:

- 18S rRNA. One of these molecules, along with some 30 different protein molecules, is used to make the small subunit of the ribosome.

- 28S, 5.8S, and 5S rRNA. One each of these molecules, along with some 45 different proteins, are used to make the large subunit of the ribosome.

The S number given each type of rRNA reflects the rate at which the molecules sediment in the ultracentrifuge. The larger the number, the larger the molecule (but not proportionally).

The 28S, 18S, and 5.8S molecules are produced by the processing of a single primary transcript from a cluster of identical copies of a single gene. The 5S molecules are produced from a different cluster of identical genes.

Transfer RNA (tRNA)

There are some 32 different kinds of tRNA in a typical eukaryotic cell.

- Each is the product of a separate gene.
- They are small (~4S), containing 73-93 nucleotides.
- Many of the bases in the chain pair with each other forming sections of double helix.
- The unpaired regions form 3 loops.
- Each kind of tRNA carries (at its 32 end) one of the 20 amino acids (thus most amino acids have more than one tRNA responsible for them).
- At one loop, 3 unpaired bases form an anticodon.
- Base pairing between the anticodon and the complementary codon on a mRNA molecule brings the correct amino acid into the growing polypeptide chain.

Small Nuclear RNA (snRNA)

Approximately a dozen different genes for snRNAs, each present in multiple copies, have been identified. The snRNAs have various roles in the processing of the other classes of RNA.

For example, several snRNAs are part of the spliceosomes that participate in converting pre-mRNA into mRNA by excising the introns and splicing the exons. [Link down to the discussion of RNA processing.]

Small Nucleolar RNA (snoRNA)

As the name suggests, these small (60–300 nucleotides) RNAs are found in the nucleolus where they are responsible for several functions:

- Some participate in making ribosomes by helping to cut up the large RNA precursor of the 28S, 18S, and 5.8S molecules.
- Others chemically modify many of the nucleotides in rRNA, tRNA, and snRNA molecules, e.g., by adding methyl groups to ribose.
- Some have been implicated in the alternative splicing of pre-mRNA to different forms of mature mRNA.
- One snoRNA serves as the template for the synthesis of telomeres.

In vertebrates, the snoRNAs are made from introns removed during RNA processing.

Noncoding RNA

Only messenger RNA encodes polypeptides. All the other classes of RNA, including types not mentioned here, are thus called noncoding RNA. Much remains to be learned about the function(s) of some of them. But, taken together, noncoding RNAs probably account for two-thirds of the transcription going on in the nucleus.

The RNA Polymerases

The RNA polymerases are huge multi-subunit protein complexes. Three kinds are found in eukaryotes.

- RNA polymerase I (Pol I). It transcribes the rRNA genes for the precursor of the 28S, 18S, and 5.8S molecules (and is the busiest of the RNA polymerases).
- RNA polymerase II (Pol II; also known as RNAP II). It transcribes protein-encoding genes into mRNA (and also the snRNA genes).
- RNA polymerase III (Pol III). It transcribes the 5S rRNA genes and all the tRNA genes.

Two remarkable reports in the 8 June 2001 issue of Science show the structure of Pol II and reveal many details of how it uses the antisense strand of DNA to synthesize a strand of mRNA.

RNA Processing: Pre-mRNA '! mRNA

All the primary transcripts produced in the nucleus must undergo processing steps to produce functional RNA molecules for export to the cytosol. We shall confine ourselves to a view of the steps as they occur in the processing of pre-mRNA to mRNA. Most eukaryotic genes are split into segments. In decoding the open reading frame of a gene for a known protein, one usually encounters periodic stretches of DNA calling for amino acids that do not occur in the actual protein product of that gene. Such stretches of DNA, which get transcribed into RNA but not translated into protein, are called introns. Those stretches of DNA that do code for amino acids in the protein are called exons. Examples:

- The gene for one type of collagen found in chickens is split into 52 separate exons.
- The gene for dystrophin, which is mutated in boys with muscular dystrophy, has 79 exons.
- Even the genes for rRNA and tRNA are split by introns.
- The human genome is estimated to contain some 180,000 exons. With a current estimate of 21,000 genes, the average exon content of our genes is about 9.

In general, introns tend to be much longer than exons. An average eukaryotic exon is only 140 nucleotides long, but one human intron stretches for 480,000 nucleotides!

Removal of the introns — and splicing the exons together — are among the essential steps in synthesizing mRNA.

The Steps of RNA Processing

- Synthesis of the cap. This is a modified guanine (G) which is attached to the 52 end of the pre-mRNA as it emerges from RNA polymerase II (RNAP II). The cap
  - protects the RNA from being degraded by enzymes that degrade RNA from the 52 end;
  - serves as an assembly point for the proteins needed to recruit the small subunit of the ribosome to begin translation.
- Step-by-step removal of introns present in the pre-mRNA and splicing of the remaining exons. This step takes place as the pre-mRNA continues to emerge from RNAP II.
- Synthesis of the poly(A) tail. This is a stretch of adenine (A) nucleotides. When a special poly(A) attachment site in the pre-mRNA emerges from RNAP II, the transcript is cut there, and the poly(A) tail is attached to the exposed 32 end. This completes the mRNA molecule, which is now ready for export to the

cytosol. (The remainder of the transcript is degraded, and the RNA polymerase leaves the DNA).

The cutting and splicing of mRNA must be done with great precision. If even one nucleotide is left over from an intron or one is removed from an exon, the reading frame from that point on will be shifted, producing new codons specifying a totally different sequence of amino acids from that point to the end of the molecule (which often ends prematurely anyway when the shifted reading frame generates a STOP codon).

The removal of introns and splicing of exons is done by spliceosomes. These are a complexes of 5 snRNA molecules and some 145 different proteins.

The introns in most pre-mRNAs begin with a GU and end with an AG. Presumably these short sequences assist in guiding the spliceosome.

Visual Evidence

The upper image is an electron micrograph of a mRNA-DNA hybrid molecule formed by mixing the messenger RNA (mRNA) from a clone of antibody-secreting cells with single-stranded DNA from the same kind of cells. The bar represents the length of 1000 bases.

The lower diagram is an interpretation of the micrograph. The solid line represents the DNA; the dotted line the mRNA. The loops (I_A, I_B, etc.) represent the introns that separate the exons encoding the domains of an antibody heavy chain:

- V = variable region
- E_1 = first constant region (C_H1) domain
- E_H = hinge region
- E_2 and E_3 = the nucleotides encoding the two C-terminal domains (C_H2 and C_H3).

The unhybridized portion of the mRNA is its poly(A) tail.

Alternative Splicing

The processing of pre-mRNA for many proteins proceeds along various paths in different cells or under different conditions. For example, early in the differentiation of a B cell (a lymphocyte that synthesizes an antibody) the cell first uses an exon that encodes a transmembrane domain that causes the molecule to be retained at the cell surface. Later, the B cell switches to using a different exon whose domain enables the protein to be secreted from the cell as a circulating antibody molecule.

Alternative splicing provides a mechanism for producing a wide variety of proteins from a small number of genes. While we humans may turn out to have only some 20 thousand genes, we probably make at least 10 times that number of different proteins. It is now estimated that 92–94% of our genes produce pre-mRNAs that are alternatively-spliced. There is evidence that the pattern of alternative splicing differs consistently in different tissues and so must be regulated. But whether all the products are functional or that many are simply the outcome of an error-prone process remains to be seen.

Alternative splicing not only provides different proteins from a single gene but also different 3' UTRs and 5' UTRs. Although not translated into protein, these untranslated regions contain signals that, for example, dictate where in the cell that protein will accumulate. Two examples:

- The 3' UTR of the *bicoid* gene in Drosophila directs the mRNA to the anterior of the embryo;
- the same region in the *VegT* gene of Xenopus directs its mRNA to the vegetal pole of the embryo.

One of the most dramatic examples of alternative splicing is the *Dscam* gene in Drosophila. This single gene contains some 116 exons of which 17 are retained in the final mRNA. Some exons are always included; others are selected from an array. Theoretically this system is able to produce 38,016 different proteins. And, in fact, over 18,000 different ones have been found in Drosophila hemolymph.

These Dscam proteins are used to establish a unique identity for each neuron. It works like this. Each developing neuron draws upon the pool of thousands of possible different mRNAs to synthesize a dozen or so of them. Which ones are selected appears to be simply a matter of chance, but because of the size of the pool, each neuron will most likely end up with a unique set of a dozen or so Dscam proteins. As each developing neuron in the central nervous system sprouts dendrites and an axon, these express its unique collection of Dscam proteins. If the extensions of a single neuron should meet each other in the tangled web that is the hallmark of nervous tissue, they are repelled. In this way, thousands of different neurons can coexist in intimate contact without the danger of nonfunctional contacts between the various extensions of the same neuron.

We also have a *DSCAM* gene which is located on chromosome 21. When present in 3 copies, it is probably responsible for some of the features of Down syndrome (and accounts for its full name: Down Syndrome Cell Adhesion Molecule). Perhaps the incredible diversity of synaptic junctions in the mammalian c.n.s. ($\sim 10^{14}$) is mediated by alternative splicing of a limited number of gene transcripts.

So, whether a particular segment of RNA will be retained as an exon or excised as an intron can vary under different circumstances, such as;

- what type of cell the gene is in;
- what stage of differentiation that cell is passing through;
- what extracellular signals that cell is receiving.

Clearly the switching to an alternate splicing pathway must be closely regulated.

Trans-splicing

Most genes are transcribed and their transcripts processed as described above. RNA polymerase travels down a single strand of a singe gene locus to form pre-mRNA that is processed (including removal of introns) to form the mature mRNA. But there are exceptions. A number of cases have been found where two different precursor transcripts have been spliced together to form the final RNA molecule. The phenomenon is called *trans*-splicing.

Examples: synthesis of a single RNA molecule by splicing together transcripts from ;

- loci located far apart on the same chromosome;
- opposite strands of the same gene locus;
- the two alleles of the gene on their separate (homologous) chromosomes.

The biological importance of these *trans*-spliced transcripts is still unknown for most of them.

Why Split Genes?

Perhaps during evolution, eukaryotic genes have been assembled from smaller, primitive genes - today's exons. Some proteins, like the antibodies mentioned in the previous section, are organized in a set of separate sections or domains each with a special function to perform in the complete molecule. Each domain is encoded by a separate exon. Having the different functional parts of the antibody molecule encoded by separate exons makes it possible to use these units in different combinations. Thus a set of exons in the genome may be the genetic equivalent of the various modular pieces in a box of "Lego" for children to assemble in whatever forms they wish. But the boundaries of other exons do not seem to correspond to domain boundaries of the protein. So perhaps the major benefit of split genes is simply the opportunity they provide for making many different proteins from a single gene through alternative splicing.

Summary

Gene expression occurs in two steps:

- transcription of the information encoded in DNA into a molecule of RNA (described here) and
- translation of the information encoded in the nucleotides of mRNA into a defined sequence of amino acids in a protein (discussed in Gene Translation: RNA → Protein).

Messenger RNA

Messenger RNA (mRNA) is a molecule of RNA encoding a chemical "blueprint" for a protein product. mRNA is transcribed from a DNA template, and carries coding information to the sites of protein synthesis: the ribosomes. Here, the nucleic acid polymer is translated into a polymer of amino acids: a protein. In mRNA as in DNA, genetic information is encoded in the sequence of nucleotides arranged into codons consisting of three bases each. Each codon encodes for a specific amino acid, except the stop codons that terminate protein synthesis. This process requires two other types of RNA: transfer RNA (tRNA) mediates recognition of the codon and provides the corresponding amino acid, while ribosomal RNA (rRNA) is the central component of the ribosome's protein manufacturing machinery.

Synthesis, Processing, and Function

The brief existence of an mRNA molecule begins with transcription and ultimately ends in degradation. During its life, an mRNA molecule may also be processed, edited, and transported prior to translation. Eukaryotic mRNA molecules often require extensive processing and transport, while prokaryotic molecules do not.

Transcription

During transcription, RNA polymerase makes a copy of a gene from the DNA to mRNA as needed. This process is similar in eukaryotes and prokaryotes. One notable difference, however, is that prokaryotic RNA polymerase associates with mRNA processing enzymes during transcription so that processing can proceed quickly after the start of transcription. The short-lived, unprocessed or partially processed, product is termed *pre-mRNA*; once completely processed, it is termed *mature mRNA*.

Eukaryotic Pre-mRNA Processing

Processing of mRNA differs greatly among eukaryotes, bacteria and archea. Non-eukaryotic mRNA is essentially mature upon transcription and requires no processing, except in rare cases. Eukaryotic pre-mRNA, however, requires extensive processing.

5' Cap Addition

A *5' cap* (also termed an RNA cap, an RNA 7-methylguanosine cap or an RNA m⁷G cap) is a modified guanine nucleotide that has been added to the "front" or 5' end of a eukaryotic messenger RNA shortly after the start of transcription. The 5' cap consists of a terminal 7-methylguanosine residue which is linked through a 5'-5'-triphosphate bond to the first transcribed nucleotide. Its presence is critical for recognition by the ribosome and protection from RNases. Cap addition is coupled to transcription, and occurs co-transcriptionally, such that each influences the other. Shortly after the start of transcription, the 5' end of the mRNA being synthesized is bound by a cap-synthesizing complex associated with RNA polymerase. This enzymatic complex catalyzes the chemical reactions that are required for mRNA capping. Synthesis proceeds as a multi-step biochemical reaction.

Splicing

Splicing is the process by which pre-mRNA is modified to remove certain stretches of non-coding sequences called introns; the stretches that remain include protein-coding sequences and are called exons. Sometimes pre-mRNA messages may be spliced in several different ways, allowing a single gene to encode multiple proteins. This process is called alternative splicing. Splicing is usually performed by an RNA-protein complex called the spliceosome, but some RNA molecules are also capable of catalyzing their own splicing.

Ribozyme

A ribozyme (from ribonucleic acid enzyme, also called RNA enzyme or catalytic RNA) is an RNA molecule possessing a well defined tertiary structure that enables it to catalyze a chemical reaction. Many natural ribozymes catalyze either the hydrolysis of one of their own phosphodiester bonds (self-cleaving ribozymes), or the hydrolysis of bonds in other RNAs, but they have also been found to catalyze the aminotransferase activity of the ribosome.

Investigators studying the origin of life have produced ribozymes in the laboratory that are capable of catalyzing their own synthesis under very specific conditions, such as an RNA polymerase ribozyme. Mutagenesis and selection has been performed resulting in isolation of improved variants of the "Round-18" polymerase ribozyme from 2001. "B6.61" is able to add up to 20 nucleotides to a primer template in 24 hours, until it decomposes by hydrolysis of its phosphodiester bonds.

Some ribozymes may play an important role as therapeutic agents, as enzymes which tailor defined RNA sequences, as biosensors, and for applications in functional genomics and gene discovery.

Discovery

Before the discovery of ribozymes, enzymes, which are defined as catalytic proteins, were the only known biological catalysts. In 1967, Carl Woese, Francis Crick, and Leslie Orgel were the first to suggest that RNA could act as a catalyst. This idea was based upon the discovery that RNA can form complex secondary structures. The first ribozymes were discovered in the 1980s by Thomas R. Cech, who was studying RNA splicing in the ciliated protozoan *Tetrahymena thermophila* and Sidney Altman, who was working on the bacterial RNase P complex.

These ribozymes were found in the intron of an RNA transcript, which removed itself from the transcript, as well as in the RNA component of the RNase P complex, which is involved in the maturation of pre-tRNAs. In 1989, Thomas R. Cech and Sidney Altman won the Nobel Prize in chemistry for their "discovery of catalytic properties of RNA." The term *ribozyme* was first introduced by Kelly Kruger *et al.* in 1982 in a paper published in *Cell*.

It had been a firmly established belief in biology that catalysis was reserved for proteins. In retrospect, catalytic RNA makes a lot of sense. This is based on the old question regarding the origin of life: Which comes first, enzymes that do the work of the cell or nucleic acids that carry the information required to produce the enzymes? Nucleic acids as catalysts circumvents this problem.

In the 1970s Thomas Cech, at the University of Colorado at Boulder, was studying the excision of introns in a ribosomal RNA gene in *Tetrahymena thermophila*. While trying to purify the enzyme responsible for splicing reaction, he found that intron could be spliced out in the absence of any added cell extract. As much as they tried, Cech and his colleagues could not identify any protein associated with the splicing reaction. After much work, Cech proposed that the intron sequence portion of the RNA could break and reform phosphodiester bonds. At about the same time, Sidney Altman, who is a Professor at Yale University, was studying the way tRNA molecules are processed in the cell when he and his colleagues isolated an enzyme called RNase-P, which is responsible for conversion of a precursor tRNA into the active tRNA.

Much to their surprise, they found that RNase-P contained RNA in addition to protein and that RNA was an essential component of the active enzyme. This was such a foreign idea that they had difficulty publishing their findings. The following year, Altman demonstrated that RNA can act as a catalyst by showing that the RNase-P RNA subunit could catalyze the cleavage of precursor tRNA into active tRNA in the absence of any protein component.

Since Cech's and Altman's discovery, other investigators have discovered other examples of self-cleaving RNA or catalytic RNA molecules. Many ribozymes have either a hairpin – or hammerhead – shaped active center and a unique secondary structure that allows them to cleave other RNA molecules at specific sequences. It is now possible to make ribozymes that will specifically cleave any RNA molecule. These RNA catalysts may have pharmaceutical applications. For example, a ribozyme has been designed to cleave the RNA of HIV. If such a ribozyme was made by a cell, all incoming virus particles would have their RNA genome cleaved by the ribozyme, which would prevent infection.

Activity

Although most ribozymes are quite rare in the cell, their roles are sometimes essential to life. For example, the functional part of the ribosome, the molecular machine that translates RNA into proteins, is fundamentally a ribozyme, composed of RNA tertiary structural motifs that are often coordinated to metal ions such as Mg^{2+} as cofactors. There is no requirement for divalent cations in a five-nucleotide RNA that can catalyze *trans*-phenylalanation of a four-nucleotide substrate which has three base complementary sequence with the catalyst. The catalyst and substrate were devised by truncation of the C3 ribozyme.

RNA can also act as a hereditary molecule, which encouraged Walter Gilbert to propose that in the distant past, the cell used RNA as both the genetic material and the structural and catalytic molecule, rather than dividing these functions between DNA and protein as they are today. This hypothesis became known as the "RNA world hypothesis" of the origin of life.

If ribozymes were the first molecular machines used by early life, then today's remaining ribozymes — such as the ribosome machinery — could be considered living fossils of a life based primarily on nucleic acids. A recent test-tube study of prion folding suggests that an RNA may catalyze the pathological protein conformation in the manner of a chaperone enzyme.

Known Ribozymes

Naturally occurring ribozymes include:

- Peptidyl transferase 23S rRNA
- RNase P
- Group I and Group II introns
- GIR1 branching ribozyme
- Leadzyme - Although initially created *in vitro*, natural examples have been found
- Hairpin ribozyme
- Hammerhead ribozyme
- HDV ribozyme
- Mammalian CPEB3 ribozyme
- VS ribozyme
- *glmS* ribozyme
- CoTC ribozyme.

Artificial Ribozymes

Since the discovery of ribozymes that exist in living organisms, there has been interest in the study of new synthetic ribozymes made in the laboratory. For example, artificially-produced self-cleaving RNAs that have good enzymatic activity have been produced. Tang and Breaker isolated self-cleaving RNAs by in vitro selection of RNAs originating from random-sequence RNAs. Some of the synthetic ribozymes that were produced had novel structures, while some were similar to the naturally occurring hammerhead ribozyme. The techniques used to discover artificial ribozymes involve Darwinian evolution. This approach takes advantage of RNA's dual nature as both a catalyst and an informational polymer, making it easy for an investigator to produce vast populations of RNA catalysts using polymerase enzymes. The ribozymes are mutated by reverse transcribing them with reverse transcriptase into various cDNA and amplified with mutagenic PCR. The selection parameters in these experiments often differ. One approach for selecting a ligase ribozyme involves using biotin tags, which are covalently linked to the substrate. If a molecule possesses the desired ligase activity, a streptavidin matrix can be used to recover the active molecules.

Lincoln and Joyce developed an RNA enzyme system capable of self replication in about an hour. By utilizing molecular competition (*in vitro* evolution) of a candidate enzyme mixture, a pair of RNA enzymes emerged, in which each synthesizes the other from synthetic oligonucleotides, with no protein present.

Applications

A type of synthetic ribozyme directed against HIV RNA called gene shears has been developed and has entered clinical testing for HIV infection.

Editing

In some instances, an mRNA will be edited, changing the nucleotide composition of that mRNA. An example in humans is the apolipoprotein B mRNA, which is edited in some tissues, but not others. The editing creates an early stop codon, which upon translation, produces a shorter protein.

Polyadenylation

Polyadenylation is the addition of a poly(A) tail to an RNA molecule. The poly(A) tail consists of multiple adenosine monophosphates; in other words, it is a stretch of RNA which only has adenine bases. In eukaryotes, polyadenylation is part of the process that produces mature messenger RNA (mRNA) for translation. It therefore forms part of the larger process of gene expression.

The process of polyadenylation begins as the transcription of a gene finishes. The 3'-most segment of the newly-made RNA is first cleaved off by a set of proteins; these proteins then synthesize the poly(A) tail at the RNA's 3' end. In some genes these proteins may add a poly(A) tail at any one of several possible sites, polyadenylation can therefore produce more than one transcript from a single gene, similar to alternative splicing.

The poly(A) tail is important for the nuclear export, translation and stability of mRNA. The tail is shortened over time and when it is short enough, the mRNA is enzymatically degraded. However, in a few cell types, mRNAs with short poly(A) tails are stored for later activation by re-polyadenylation in the cytosol. In contrast, when polyadenylation occurs in bacteria, it promotes RNA degradation. This is also sometimes the case for eukaryotic non-coding RNAs. The wide distribution of polyadenylation among living organisms indicates that this process evolved early in the history of life on Earth.

Primer on RNA

RNAs are a type of large biological molecules, whose individual building blocks are called nucleotides. The name *poly(A) tail* (for polyadenylic acid tail) reflects the way RNA nucleotides are abbreviated, with a letter for the base the nucleotide contains (A for adenine, C for cytosine, G for guanine and U for uracil). RNAs are produced (*transcribed*) from a DNA template. By convention, RNA sequences are written in a 5' to 3' direction. The 5' end is the part of the RNA molecule that is transcribed first, and the 3' end is transcribed last. The 3' end is also where the poly(A) tail is found on polyadenylated RNAs.

Messenger RNA (mRNA) is RNA that has a coding region that acts as a template for protein synthesis (*translation*). The rest of the mRNA, the *untranslated regions*, tune how active the mRNA is. There are also many RNAs that are not translated, called non-coding RNAs. Like the untranslated regions, many of these non-coding RNAs have regulatory roles.

Nuclear Polyadenylation

Function

In nuclear polyadenylation, a poly(A) tail is added to an RNA at the end of transcription. On mRNAs, the poly(A) tail protects the mRNA molecule from enzymatic degradation in the cytoplasm and aids in transcription termination, export of the mRNA from the nucleus, and translation. Almost all eukaryotic mRNAs are polyadenylated, with the exception of animal replication-dependent histone mRNAs. These are the only mRNAs in eukaryotes that lack a poly(A) tail, ending instead in a stem-loop structure followed by a purine-rich sequence, termed histone downstream element, that directs where the RNA is cut so that the 3' end of the histone mRNA is formed.

Many eukaryotic non-coding RNAs are always polyadenylated at the end of transcription. There are small RNAs where the poly(A) tail is only seen in intermediary forms and not in the mature RNA as the ends are removed during processing, notably microRNAs. But for many long noncoding RNAs – a seemingly large group of regulatory RNAs that for example includes the RNA Xist which mediates X chromosome inactivation – a poly(A) tail is part of the mature RNA.

Mechanism

The polyadenylation machinery in the nucleus of eukaryotes works on products of RNA polymerase II, such as precursor mRNA. Here, a multi-protein complex cleaves the 3'-most part of a newly produced RNA and polyadenylates the end produced by this cleavage. The cleavage is catalysed by the enzyme CPSF and occurs 10–30 nucleotides downstream of its binding site.

This site is often the sequence AAUAAA on the RNA, but variants of it exist that bind more weakly to CPSF. Two other proteins add specificity to the binding to an RNA: CstF and CFI. CstF binds to a GU-rich region further downstream of CPSF's site. CFI recognises a third site on the RNA (a set of UGUAA sequences in mammals) and can recruit CPSF even if the AAUAAA sequence is missing. The polyadenylation signal – the sequence motif recognised by the RNA cleavage complex – varies between groups of eukaryotes. Most human polyadenylation sites contain the AAUAAA sequence, but this sequence is less common in plants and fungi.

The RNA is cleaved right after transcription, as CstF also binds to RNA polymerase II. Cleavage also involves the protein CFII, though it is unknown how. The cleavage site associated with a polyadenylation signal can vary up to some 50 nucleotides. When the RNA is cleaved, polyadenylation starts, catalysed by polyadenylate polymerase. Polyadenylate polymerase builds the poly(A) tail by adding adenosine monophosphate units from adenosine triphosphate to the RNA, cleaving off pyrophosphate. Another protein, PAB2, binds to the new, short poly(A) tail and increases the affinity of polyadenylate polymerase for the RNA. When the poly(A) tail is approximately 250 nucleotides long the enzyme can no longer bind to CPSF and polyadenylation stops, thus determining the length of the poly(A) tail. CPSF is in contact with RNA polymerase II, allowing it to tell the polymerase to terminate transcription. The polyadenylation machinery is also physically linked to the spliceosome, a complex that removes introns from RNAs.

Downstream Effects

The poly(A) tail acts as the binding site for poly(A)-binding protein. Poly(A)-binding protein promotes export from the nucleus and translation, and inhibits degradation. This protein binds to the poly(A) tail prior to mRNA export from the nucleus and in yeast also recruits poly(A) nuclease, an enzyme that shortens the poly(A) tail and allows the export of the mRNA. Poly(A)-binding protein is exported to the cytoplasm with the RNA. mRNAs that are not exported are degraded by the exosome. Poly(A)-binding protein also can bind to, and thus recruit, several proteins that affect translation, one of these is initiation factor-4G which in turn recruits the 40S ribosomal subunit. However, a poly(A) tail is not required for the translation of all mRNAs.

Deadenylation

In eukaryotic somatic cells, the poly(A) tail of most mRNAs in the cytoplasm gradually get shorter, and mRNAs with shorter poly(A) tail are translated less and degraded sooner. However, it can take many hours before an mRNA is degraded. This deadenylation and degradation process can be accelerated by microRNAs complementary to the 3' untranslated region of an mRNA. In immature egg cells, mRNAs with shortened poly(A) tails are not degraded, but are instead stored without being translated. They are then activated by cytoplasmic polyadenylation after fertilisation, during egg activation.

In animals, poly(A) ribonuclease (PARN) can bind to the 5' cap and remove nucleotides from the poly(A) tail. The level of access to the 5' cap and poly(A) tail is important in controlling how soon the mRNA is degraded. PARN deadenylates less if the RNA is bound by the initiation factors 4E (at the 5' cap) and 4G (at the poly(A) tail), so this is why translation reduces deadenylation. The rate of deadenylation may also be regulated by RNA-binding proteins. Once the poly(A) tail is removed, the decapping complex removes the 5' cap, leading to a degradation of the RNA. Several other enzymes that seem to be involved in deadenylation have been identified in yeast.

Alternative Polyadenylation

Many protein-coding genes have more than one polyadenylation site, so a gene can code for several mRNAs that differ in their 3' end. Since alternative polyadenylation changes the length of the 3' untranslated region, it can change which binding sites for microRNAs the 3' untranslated region contains. MicroRNAs tend to repress translation and promote degradation of the mRNAs they bind to, although there are examples of microRNAs that stabilise transcripts. Alternative polyadenylation can also shorten the coding region, thus making the mRNA code for a different protein, but this is much less common than just shortening the 3' untranslated region.

The choice of poly(A) site can be influenced by extracellular stimuli and depends on the expression of the proteins that take part in polyadenylation. For example, the expression of CstF-64, a subunit of cleavage stimulatory factor (CstF), increases in macrophages in response to lipopolysaccharides (a group of bacterial compounds that trigger an immune response). This results in the selection of weak poly(A) sites and thus shorter transcripts. This removes regulatory elements in the 3' untranslated regions of mRNAs for defense-related products like lysozyme and TNF-α. These mRNAs then have longer half-lives and produce more of these proteins. RNA-binding proteins other than those in the polyadenylation machinery can also affect whether a polyadenyation site is used,as can DNA methylation near the polyadenylation signal.

Cytoplasmic Polyadenylation

There is polyadenylation in the cytosol of some animal cell types, namely in the germ line, during early embryogenesis and in post-synaptic sites of nerve cells. This lengthens the poly(A) tail of an mRNA with a shortened poly(A) tail, so that the mRNA will be translated. These shortened poly(A) tails are often less than 20 nucleotides, and are lengthened to around 80–150 nucleotides.

In the early mouse embryo, cytoplasmic polyadenylation of maternal RNAs from the egg cell allows the cell to survive and grow even though transcription does not start until the middle of the 2-cell stage (4-cell stage in human). In the brain, cytoplasmic polyadenylation is active during learning and could play a role in long-term potentiation, which is the strengthening of the signal transmission from a nerve cell to another in response to nerve impulses and is important for learning and memory formation.

Cytoplasmic polyadenylation requires the RNA-binding proteins CPSF and CPEB, and can involve other RNA-binding proteins like Pumilio. Depending on the cell type, the polymerase can be the same type of polyadenylate polymerase (PAP) that is used in the nuclear process, or the cytoplasmic polymerase GLD-2.

Tagging for Degradation in Eukaryotes

For many non-coding RNAs, including tRNA, rRNA, snRNA and snoRNA, polyadenylation is a way of marking the RNA for degradation, in at least yeast. This polyadenylation is done in the nucleus by the TRAMP complex, which adds a tail that is around 40 nucleotides long to the 3' end. The RNA is then degraded by the exosome. Poly(A) tails have also been found on human rRNA fragments, both the form of homopolymeric (A only) and heterpolymeric (mostly A) tails.

In Prokaryotes and Organelles

In many bacteria, both mRNAs and non-coding RNAs can be polyadenylated. This poly(A) tail promotes degradation by the degradosome, which contains two RNA-degrading enzymes: polynucleotide phosphorylase and RNase E. Polynucleotide phosphorylase binds to the 3' end of RNAs and the 3' extension provided by the poly(A) tail allows it to bind to the RNAs whose secondary structure would otherwise block the 3' end. Successive rounds of polyadenylation and degradation of the 3' end by polynucleotide phosphorylase allows the degradosome to overcome these secondary structures. The poly(A) tail can also recruit RNases that cut the RNA in two. These bacterial poly(A) tails are about 30 nucleotides long.

In as different groups as animals and trypanosomes, the mitochondria contain both stabilising and destabilising poly(A) tails. Destabilising polyadenylation targets both mRNA and noncoding RNAs. The poly(A) tails are 43 nucleotides long on average. The stabilising ones start at the stop codon, and without them the stop codon (UAA) is not complete as the genome only encodes the U or UA part. Plant mitochondria only have destabilising polyadenylation, and yeast mitochondria have no polyadenylation at all.

While many bacteria and mitochondria have polyadenylate polymerases, they also have another type of polyadenylation, performed by polynucleotide phosphorylase itself. This enzyme is found in bacteria, mitochondria, plastids and as a constituent of the archeal exosome (in those archaea that have an exosome). It can synthesise a 3' extension where the vast majority of the bases are adenines. Like in bacteria, polyadenylation by polynucleotide phosphorylase promotes degradation of the RNA in plastids and likely also archaea.

Evolution

Although polyadenylation is seen in almost all organisms, it is not universal. However, the wide distribution of this modification and the fact that it is present in organisms from all three domains of life implies that the last universal common ancestor of all living organisms probably had some form of polyadenylation system. A few organisms do not polyadenylate mRNA, which implies that they have lost their polyadenylation machineries during evolution. Although no examples of eukaryotes that lack polyadenylation are known, mRNAs from the bacterium *Mycoplasma gallisepticum* and the salt-tolerant archaean *Haloferax volcanii* lack this modification.

The most ancient polyadenylating enzyme is polynucleotide phosphorylase. This enzyme is part of both the bacterial degradosome and archaeal exosome, two closely related complexes that recycle RNA into nucleotides. This enzyme degrades RNA by attacking the bond between the 3'-most nucleotides with a phosphate, breaking off a diphosphate nucleotide.

This reaction is reversible, and so the enzyme can also extend RNA with more nucleotides. The heteropolymeric tail added by polynucleotide phosphorylase is very rich in adenine. The choice of adenine is most likely the result of higher ADP concentrations than other nucleotides as a result of using ATP as an energy currency, making it more likely to be incorporated in this tail in early lifeforms. It has been suggested that the involvement of adenine-rich tails in RNA degradation prompted the later evolution of polyadenylate polymerases (the enzymes that produce poly(A) tails with no other nucleotides in them).

Polyadenylate polymerases are not as ancient. They have separately evolved in both bacteria and eukaryotes from CCA-adding enzyme, which is the enzyme that completes the 3' ends of tRNAs. Its catalytic domain is homologous to that of other polymerases. Likely, the horizontal transfer of bacterial CCA-adding enzyme to eukaryotes allowed the archaeal-like CCA-adding enzyme to switch function to a poly(A) polymerase. Some lineages, like archaea and cyanobacteria, never evolved a polyadenylate polymerase.

History

Polyadenylation was first identified in 1960 as an enzymatic activity in extracts made from cell nuclei that could polymerise ATP, but not ADP, into polyadenine. Although identified in many types of cells, this activity had no known function until 1971, when poly(A) sequences were found in mRNAs. Initially, the only function of these sequences was thought to be protection of the 3' end of the RNA from nucleases, but later the specific roles of polyadenylation in nuclear export and translation were identified. The polymerases responsible for polyadenylation were first purified and characterized in the 1960s and 1970s, but the large number of accessory proteins that control this process were only discovered in the early 1990s.

Polyadenylation is the covalent linkage of a polyadenylyl moiety to a messenger RNA molecule. In eukaryotic organisms, most messenger RNA (mRNA) molecules are polyadenylated at the 3' end. The poly(A) tail and the protein bound to it aid in protecting mRNA from degradation by exonucleases. Polyadenylation is also important for transcription termination, export of the mRNA from the nucleus, and translation. mRNA can also be polyadenylated in prokaryotic organisms, where poly(A) tails act to facilitate, rather than impede, exonucleolytic degradation.

Polyadenylation occurs during and immediately after transcription of DNA into RNA. After transcription has been terminated, the mRNA chain is cleaved through the action of an endonuclease complex associated with RNA polymerase. After the mRNA has been cleaved, around 250 adenosine residues are added to the free 3' end at the cleavage site. This reaction is catalyzed by polyadenylate polymerase. Just as in alternative splicing, there can be more than one polyadenylation variant of a mRNA.

Transport

Another difference between eukaryotes and prokaryotes is mRNA transport. Because eukaryotic transcription and translation is compartmentally separated, eukaryotic mRNAs must be exported from the nucleus to the cytoplasm. Mature mRNAs are recognized by their processed modifications and then exported through the nuclear pore. In neurons mRNA must be transported from the soma to the dendrites where local translation occurs in response to external stimuli. Many messages are marked with so-called "zip codes" which targets their transport to a specific location.

Translation

Because prokaryotic mRNA does not need to be processed or transported, translation by the ribosome can begin immediately after the end of transcription. Therefore, it can be said that prokaryotic translation is *coupled* to transcription and occurs *co-transcriptionally*.

Eukaryotic mRNA that has been processed and transported to the cytoplasm (i.e. mature mRNA) can then be translated by the ribosome. Translation may occur at ribosomes free-floating in the cytoplasm, or directed to the endoplasmic reticulum by the signal recognition particle. Therefore, unlike in prokaryotes, eukaryotic translation *is not* directly coupled to transcription.

Structure

5' cap

The 5' cap is a specially altered nucleotide on the 5' end of precursor messenger RNA and some other primary RNA transcripts as found in eukaryotes. The process of 5' capping is vital to creating mature messenger RNA, which is then able to undergo translation. Capping ensures the messenger RNA's stability while it undergoes translation in the process of protein synthesis, and is a highly regulated process that occurs in the cell nucleus.

5' Cap Structure

The 5' cap is found on the 5' end of an mRNA molecule and consists of a guanine nucleotide connected to the mRNA via an unusual 5' to 5' triphosphate linkage. This guanosine is methylated on the 7 position directly after capping *in vivo* by a methyl transferase. It is referred to as a 7-methylguanosine cap, abbreviated m^7G.

Further modifications include the possible methylation of the 2' hydroxy-groups of the first 2 ribose sugars of the 5' end of the mRNA. The methylation of both 2' hydroxy-groups is shown on the diagram.

The 5' cap looks like the 3' end of an RNA molecule (the 5' carbon of the cap ribose is bonded, and the 3' unbonded). This provides significant resistance to 5' exonucleases.

Capping Process

The starting point is the unaltered 5' end of an RNA molecule. This features a final nucleotide followed by three phosphate groups attached to the 5' carbon.

1. One of the terminal phosphate groups is removed (by RNA terminal phosphatase), leaving two terminal phosphates.

2. GTP is added to the terminal phosphates (by a guanylyl transferase), losing two phosphate groups (from the GTP) in the process. This results in the 5' to 5' triphosphate linkage.

3. The 7-nitrogen of guanine is methylated (by a methyl transferase).

4. Other methyltransferases are optionally used to carry out methylation of 5' proximal nucleotides.

5' Capping Targeting

The Capping Enzyme Complex (CEC) required for capping is found bound to the RNA polymerase II before transcription starts. As soon as the 5' end of the new transcript emerges the enzymes transfer to it and begin the capping process (this is a similar kind of mechanism to ensure capping as for polyadenylation).

The enzymes for capping can only bind to RNA polymerase II ensuring specificity to only these transcripts, which are almost entirely mRNA.

5' Cap Function

The 5' cap has 4 main functions:

1. Regulation of nuclear export.

2. Prevention of degradation by exonucleases.

3. Promotion of translation.

4. Promotion of 5' proximal intron excision.

Nuclear export of RNA is regulated by the Cap binding complex (CBC), which binds exclusively to capped RNA. The CBC is then recognized by the nuclear pore complex and exported. Once in the cytoplasm after the pioneer round of translation, the CBC is replaced by the translation factors eIF-4E and eIF-4G. This complex is then recognized by other translation initiation machinery including the ribosome.

Cap prevents 5' degradation in two ways. First, degradation of the mRNA by 5' exonucleases is prevented (as mentioned above) by functionally looking like a 3' end. Second, the CBC complex and the eIF-4E/eIF-4G block the access of decapping enzymes to the cap. This increases the half-life of the mRNA, essential in eukaryotes as the export process takes significant time.

Decapping of an mRNA is catalyzed by the decapping complex made up of at least Dcp1 and Dcp2, which must compete with eIF-4E to bind the cap. Thus the 5' cap is a marker of an actively translating mRNA and is used by cells to regulate mRNA half-lives in response to new stimuli. Undesirable mRNAs are sent to P-bodies for temporary storage or decapping, the details of which are still being resolved.

The mechanism of 5' proximal intron excision promotion is not well understood, but the 5' cap appears to loop around and interact with the spliceosome in the splicing process, promoting intron excision.

Messenger RNA Decapping

The process of messenger RNA decapping consists of hydrolysis of the 5' cap structure on the RNA exposing a 5' monophosphate. This 5' monophosphate is a substrate for the exonuclease Xrn1 and the message is quickly destroyed. There are many situations which may lead to the removal of the cap, some of which are discussed below.

Translation and Decay

Inside cells there is a balance between the processes of translation and mRNA decay. Messages which are being actively translated are bound by polysomes and the initiation factors eIF-4E and eIF-4G. This blocks access to the cap by the decapping enzyme DCP2 and protects the message. In nutrient-starvation conditions or viral infection, translation may be compromised and decapping is stimulated. This balance is reflected in the size and abundance of the cytoplasmic structures known as P-bodies

Specific Decay Pathways

A number of specific decay pathways exist that recognize aberrant messages and promote their decapping. Nonsense mediated decay recognizes premature stop codons and promotes decapping as well as decay by the exosome. Certain classes of miRNA have also been shown to stimulate decapping.

Coding Regions

The coding region of a gene is that portion of a gene's DNA or RNA, composed of exons, that codes for protein. The region is bounded nearer the 5' end by a start codon and nearer the 3' end with a stop codon. The coding region in mRNA is bounded by the five prime untranslated region and the three prime untranslated region, which are also parts of the exons.

The coding region of an organism is sum total of the organism's genome that is composed of gene coding regions. Coding regions are composed of codons, which are decoded and translated (in eukaryotes usually into one and in prokaryotes usually into several) proteins by the ribosome. Coding regions begin with the start codon and end with a stop codon. Generally, the start codon is an AUG triplet and the stop codon is UAA, UAG, or UGA. The coding regions tend to be stabilised by internal base pairs, this impedes degradation. In addition to being protein-coding, portions of coding regions may serve as regulatory sequences in the pre-mRNA as exonic splicing enhancers or exonic splicing silencers.

Open Reading Frame

In molecular genetics, an open reading frame (ORF) is a DNA sequence that does not contain a stop codon in a given reading frame.

Significance

One common use of open reading frames are as one piece of evidence to assist in gene prediction. Long ORFs are often used, along with other evidence, to initially identify candidate protein coding regions in a DNA sequence. The presence of an ORF does not necessarily mean that the region is ever translated. For example in a randomly generated DNA sequence with an equal percentage of each nucleotide, a stop-codon would be expected once every 21 codons. A simple gene prediction algorithm for prokaryotes might look for a start codon followed by an open reading frame that is long enough to encode a typical protein, where the codon usage of that region matches the frequency characteristic for the given organism's coding regions. By itself even a long open reading frame is not conclusive evidence for the presence of a gene.

Example

If a portion of a genome has been sequenced (e.g. 5'-ATCTAAAATGGGTGCC-3'), ORFs can be located by examining each of the three possible reading frames on each strand. In this sequence two out of three possible reading frames are entirely open, meaning that they do not contain a stop codon:

1.   ...A TCT AAA ATG GGT GCC...

2.   ...AT CTA AAA TGG GTG CC...

3.   ...ATC TAA AAT GGG TGC C...

Possible stop codons in DNA are "TGA", "TAA" and "TAG". Thus, the last reading frame in this example contains a stop codon (*TAA*), unlike the first two.

Untranslated Regions

Untranslated regions (UTRs) are sections of the mRNA before the start codon and after the stop codon that are not translated, termed the five prime untranslated region (5' UTR) and three prime untranslated region (3' UTR), respectively. These regions are transcribed with the coding region and thus are exonic as they are present in the mature mRNA. Several roles in gene expression have been attributed to the untranslated regions, including mRNA stability, mRNA localization, and translational efficiency. The ability of a UTR to perform these functions depends on the sequence of the UTR and can differ between mRNAs.

The stability of mRNAs may be controlled by the 5' UTR and/or 3' UTR due to varying affinity for RNA degrading enzymes called ribonucleases and for ancillary proteins that can promote or inhibit RNA degradation.

Translational efficiency, including sometimes the complete inhibition of translation, can be controlled by UTRs. Proteins that bind to either the 3' or 5' UTR may affect translation by influencing the ribosome's ability to bind to the mRNA. MicroRNAs bound to the 3' UTR also may affect translational efficiency or mRNA stability.

Cytoplasmic localization of mRNA is thought to be a function of the 3' UTR. Proteins that are needed in a particular region of the cell can actually be translated there; in such a case, the 3' UTR may contain sequences that allow the transcript to be localized to this region for translation.

Some of the elements contained in untranslated regions form a characteristic secondary structure when transcribed into RNA. These structural mRNA elements are involved in regulating the mRNA. Some, such as the SECIS element, are targets for proteins to bind. One class of mRNA element, the riboswitches, directly bind small molecules, changing their fold to modify levels of transcription or translation. In these cases, the mRNA regulates itself.

5' UTR

The five prime untranslated region (5' UTR), also known as the leader sequence, is a particular section of messenger RNA (mRNA) and the DNA that codes for it. At times the leader sequence can contain elements for controlling gene expression by way of attenuation. The leader sequence will be translated and the availability of the encoded amino acids will affect the transcription of the rest of the mRNA. It starts at the +1 position (where transcription begins) and ends one nucleotide before the start codon (usually AUG) of the coding region. It usually contains a ribosome binding site (RBS), in bacteria also known as the Shine Dalgarno sequence (AGGAGGU). The 5' UTR may be a hundred or more nucleotides long, and the 3' UTR may be even longer (up to several kilobases in length).

An mRNA molecule codes for a protein through translation. The mRNA also contains regions that are not translated: in eukaryotes this includes the 5' untranslated region, 3' untranslated region, 5' cap and poly-A tail.

In prokaryotic mRNA the 5' UTR is normally short. Some viruses and cellular genes have unusual long structured 5' UTRs which may have roles in gene expression.

Several regulatory sequences may be found in the 5' UTR:

- Binding sites for proteins, that may affect the mRNA's stability or translation, for example iron responsive elements, which occur in the 5' UTRs (and 3' UTRs) of a small number of eukaryotic mRNAs that regulate gene expression in response to iron.
- Regulatory elements that do not depend on proteins, such as riboswitches.
- Sequences that promote the initiation of translation.

3' UTR

The three prime untranslated region (3' UTR) is a particular section of messenger RNA (mRNA). It follows the coding region.

An mRNA molecule codes for a protein through translation. The mRNA also contains regions that are not translated. In eukaryotes these regions are the cap, 5' untranslated region, 3' untranslated region, and polyA tail.

In prokaryotes mRNA structures have some differences as do histone mRNAs. However, both have 3' UTRs.

Several regulatory sequences are found in the 3' UTR:

- A polyadenylation signal, usually *AAUAAA*, or a slight variant. This marks the site of cleavage of the transcript approximately

30 base pairs past the signal, followed by the addition of
several hundred adenine residues (poly-A tail).

- Binding sites for proteins, that may affect the mRNA's stability
  or location in the cell, like SECIS elements (which direct the
  ribosome to translate the codon UGA as selenocysteines rather
  than as a stop codon), or AU-rich elements (AREs), stretches
  consisting of mainly adenine and uracil nucleotides (which can
  either stabilize or destabilize the mRNA depending on the
  protein bound to it).
- Binding sites for miRNAs, a type of RNAi.

Poly(A) Tail

The 3' poly(A) tail is a long sequence of adenine nucleotides (often
several hundred) added to the 3' end of the pre-mRNA. This tail
promotes export from the nucleus and translation, and protects the
mRNA from degradation.

Monocistronic versus Polycistronic mRNA

An mRNA molecule is said to be monocistronic when it contains
the genetic information to translate only a single protein. This is the
case for most of the eukaryotic mRNAs.

On the other hand, polycistronic mRNA carries the information
of several genes, which are translated into several proteins. These
proteins usually have a related function and are grouped and regulated
together in an operon. Most of the mRNA found in bacteria and archea
are polycistronic. Dicistronic or bicistronic is the term used to describe
an mRNA that encodes only two proteins.

mRNA Circularization

In eukaryotes it is thought that mRNA molecules form circular
structures due to an interaction between the cap binding complex and
poly(A)-binding protein. Circularization is thought to promote recycling
of ribosomes on the same message leading to efficient translation.

Degradation

Different mRNAs within the same cell have distinct lifetimes
(stabilities). In bacterial cells, individual mRNAs can survive from
seconds to more than an hour; in mammalian cells, mRNA lifetimes
range from several minutes to days. The greater the stability of an
mRNA, the more protein may be produced from that mRNA. The
limited lifetime of mRNA enables a cell to alter protein synthesis

rapidly in response to its changing needs. There are many mechanisms that lead to the destruction of a mRNA, some of which are described below.

Prokaryotic mRNA Degradation

In prokaryotes the lifetime of mRNA is generally much shorter than in eukaryotes. Prokaryotes degrade messages by using a combination of ribonucleases, including endonucleases, 3' exonucleases, and 5' exonucleases. In some instances, small RNA molecules (sRNA) tens to hundreds of nucleotides long can stimulate the degradation of specific mRNAs by base pairing with complementary sequences and facilitating ribonuclease cleavage. It was recently shown that bacteria also have a sort of 5' cap consisting of a triphosphate on the 5' end. Removal of two of the phosphates leaves a 5' monophosphate, causing the message to be destroyed by the endonuclease RNase E.

Eukaryotic mRNA Turnover

Inside eukaryotic cells there is a balance between the processes of translation and mRNA decay. Messages that are being actively translated are bound by ribosomes, the eukaryotic initiation factors eIF-4E and eIF-4G, and poly(A)-binding protein. eIF-4E and eIF-4G block the decapping enzyme (DCP2), and poly(A)-binding protein blocks the exosome complex, protecting the ends of the message. The balance between translation and decay is reflected in the size and abundance of cytoplasmic structures known as P-bodies The poly(A) tail of the mRNA is shortened by specialized exonucleases that are targeted to specific messenger RNAs by a combination of cis-regulatory sequences on the RNA and trans-acting RNA-binding proteins. Poly(A) tail removal is thought to disrupt the circular structure of the message and destabilize the cap binding complex. The message is then subject to degradation by either the exosome complex or the decapping complex. In this way, translationally inactive messages can be destroyed quickly, while active messages remain intact.

AU-rich Element Decay

The presence of AU-rich elements in some mammalian mRNAs tends to destabilize those transcripts through the action of cellular proteins that bind these sequences and stimulate poly(A) tail removal. Loss of the poly(A) tail is thought to promote mRNA degradation by facilitating attack by both the exosome complex and the decapping complex. Rapid mRNA degradation via AU-rich elements is a critical mechanism for preventing the overproduction of potent cytokines such as tumour necrosis factor (TNF) and granulocyte-macrophage colony stimulating factor (GM-CSF). AU-rich elements also regulate the biosynthesis of proto-oncogenic transcription factors like c-Jun and c-Fos.

Nonsense Mediated Decay

Nonsense-mediated decay (NMD) is a cellular mechanism of mRNA surveillance that functions to detect nonsense mutations and prevent the expression of truncated or erroneous proteins. Following transcription, precursor mRNA undergoes an assemblage of ribonucleoprotein (RNP) components followed by regulatory pre-mRNA processing. Large average intron size in eukaryotic cells greatly increases the probability that aberrant mRNA splicing will result in the presence of a nonsense (stop) codon (UAA, UAG, UGA) somewhere within the open reading frame. NMD is triggered by exon junction complexes (EJCs) (components of the assembled RNP) that are deposited during pre-mRNA processing. While EJCs located in an open reading frame upstream of exon-exon junctions may facilitate ribosomal recruitment prior to displacement by a "pioneer" round of translation, EJCs located downstream of a nonsense codon are not displaced because the ribosome is released from the transcript before reaching it. These remaining EJCs function as tags for recruitment of UPF1 following the mRNA's transport out of the nucleus and into the cytosol where the RNA is degraded, for example by the exosome complex. NMD is not only a mechanism for degrading aberrant mRNA, however, as there are numerous examples of normal transcripts whose expression is regulated by this process including the plasticity protein Arc/Arg3.1.

Eukaryotic messages are subject to surveillance by nonsense mediated decay (NMD), which checks for the presence of premature stop codons (nonsense codons) in the message. These can arise via incomplete splicing, V(D)J recombination in the adaptive immune system, mutations in DNA, transcription errors, leaky scanning by the ribosome causing a frame shift, and other causes. Detection of a premature stop codon triggers mRNA degradation by 5' decapping, 3' poly(A) tail removal, or endonucleolytic cleavage.

Small Interfering RNA (siRNA)

Small interfering RNA (siRNA), sometimes known as short interfering RNA or silencing RNA, is a class of double-stranded RNA molecules, 20-25 nucleotides in length, that play a variety of roles in biology. The most notable role of siRNA is its involvement in the RNA interference (RNAi) pathway, where it interferes with the expression of a specific gene. In addition to its role in the RNAi pathway, siRNA also acts in RNAi-related pathways, e.g., as an antiviral mechanism or in shaping the chromatin structure of a genome; the complexity of these pathways is only now being elucidated.

siRNAs were first discovered by David Baulcombe's group at the Sainsbury Laboratory in Norwich, England, as part of post-transcriptional gene silencing (PTGS) in plants. The group published their findings in *Science* in a paper titled "A species of small antisense RNA in posttranscriptional gene silencing in plants". Shortly thereafter, in 2001, synthetic siRNAs were shown to be able to induce RNAi in mammalian cells by Thomas Tuschl, and colleagues in a paper published in *Nature*. This discovery led to a surge in interest in harnessing RNAi for biomedical research and drug development.

RNAi Induction Using siRNAs or their Biosynthetic Precursors

Transfection of an exogenous siRNA can be problematic because the gene knockdown effect is only transient, in particular, in rapidly dividing cells. One way of overcoming this challenge is to modify the siRNA in such a way as to allow it to be expressed by an appropriate vector, e.g., a plasmid.

This is done by the introduction of a loop between the two strands, thus producing a single transcript, which can be processed into a functional siRNA. Such transcription cassettes typically use an RNA polymerase III promoter (e.g., U6 or H1), which usually directs the transcription of small nuclear RNAs (snRNAs) (U6 is involved in gene splicing; H1 is the RNase component of human RNase P). It is assumed (although not known for certain) that the resulting siRNA transcript is then processed by Dicer.

RNA Activation

It has recently been found that dsRNA can also activate gene expression, a mechanism that has been termed "small RNA-induced gene activation" or RNAa. It has been shown that dsRNAs targeting gene promoters induce potent transcriptional activation of associated genes. RNAa was demonstrated in human cells using synthetic dsRNAs, termed "small activating RNAs" (saRNAs). It is currently not known whether RNAa is conserved in other organisms.

Challenges: Avoiding Nonspecific Effects

Because RNAi intersects with a number of other pathways, it is not surprising that on occasion nonspecific effects are triggered by the experimental introduction of an siRNA. When a mammalian cell encounters a double-stranded RNA such as an siRNA, it may mistake it as a viral by-product and mount an immune response. Furthermore, because structurally related microRNAs modulate gene expression largely via incomplete complementarity base pair interactions with a target mRNA, the introduction of an siRNA may cause unintended off-targeting.

Innate Immunity

Introduction of too much siRNA can result in nonspecific events due to activation of innate immune responses. Most evidence to date suggests that this is probably due to activation of the dsRNA sensor PKR, although retinoic acid-inducible gene I (RIG-I) may also be involved. The induction of cytokines via toll-like receptor 7 (TLR7) has also been described. One promising method of reducing the nonspecific effects is to convert the siRNA into a microRNA. MicroRNAs occur naturally, and by harnessing this endogenous pathway it should be possible to achieve similar gene knockdown at comparatively low concentrations of resulting siRNAs. This should minimize nonspecific effects.

Off-targeting

Off-targeting is another challenge to the use of siRNAs as a gene knockdown tool. Here, genes with incomplete complementarity are inadvertently downregulated by the siRNA (in effect, the siRNA acts as a miRNA), leading to problems in data interpretation and potential toxicity. This, however, can be partly addressed by designing appropriate control experiments, and siRNA design algorithms are currently being developed to produce siRNAs free from off-targeting. Genome-wide expression analysis, e.g., by microarray technology, can then be used to verify this and further refine the algorithms. A 2006 paper from the laboratory of Dr. Khvorova implicates 6- or 7-basepair-long stretches from position 2 onward in the siRNA matching with 3'UTR regions in off-targeted genes.

Possible Therapeutic Applications and Challenges

Given the ability to knock down, in essence, any gene of interest, RNAi via siRNAs has generated a great deal of interest in both basic and applied biology. There are an increasing number of large-scale RNAi screens that are designed to identify the important genes in various biological pathways. Because disease processes also depend on the activity of multiple genes, it is expected that in some situations turning off the activity of a gene with an siRNA could produce a therapeutic benefit.

However, applying RNAi via siRNAs to living animals, especially humans, poses many challenges. Under experiments, siRNAs show different effectiveness in different cell types in a manner as yet poorly understood: Some cells respond well to siRNAs and show a robust knockdown, whereas others show no such knockdown (even despite efficient transfection).

Phase I results of the first two therapeutic RNAi trials (indicated for age-related macular degeneration, aka AMD) reported at the end of 2005 that siRNAs are well tolerated and have suitable pharmacokinetic properties.

Proof of concept trials have indicated that Ebola-targeted siRNAs may be effective as post-exposure prophylaxis in humans, with 100% of non-human primates surviviing a lethal dose of Zaire Ebolavirus, the most lethal strain.

In metazoans, small interfering RNAs (siRNAs) processed by Dicer are incorporated into a complex known as the RNA-induced silencing complex or RISC. This complex contains an endonuclease that cleaves perfectly complementary messages to which the siRNA binds. The resulting mRNA fragments are then destroyed by exonucleases. siRNA is commonly used in laboratories to block the function of genes in cell culture. It is thought to be part of the innate immune system as a defense against double-stranded RNA viruses.

MicroRNA (miRNA)

MicroRNAs (miRNAs) are small RNAs that typically are partially complementary to sequences in metazoan messenger RNAs. Binding of a miRNA to a message can repress translation of that message and accelerate poly(A) tail removal, thereby hastening mRNA degradation. The mechanism of action of miRNAs is the subject of active research. MicroRNAs (miRNAs) are short ribonucleic acid (RNA) molecules, on average only 22 nucleotides long and are found in all eukaryotic cells, except fungi and marine plants. miRNAs are post-transcriptional regulators that bind to complementary sequences on target messenger RNA transcripts (mRNAs), usually resulting in translational repression and gene silencing. The human genome may encode over 1000 miRNAs, which may target about 60% of mammalian genes and are abundant in many human cell types.

miRNAs show very different characteristics between plants and metazoans. In plants the miRNA complementarity to its mRNA target is nearly perfect, with no or few mismatched bases. In metazoans on the other hand miRNA complementarity is far from perfect and one miRNA can target many different sites on the same mRNA or on many different mRNAs. Another difference is the location of target sites on mRNAs. In metazoans the miRNA target sites are in the three prime untranslated regions (3'UTR) of the mRNA. In plants targets can be located in the 3' UTR but are more often in the coding region itself. MiRNAs are well conserved in eukaryotic organism and are thought to be a vital and evolutionarily ancient component of genetic regulation.

The first miRNAs were characterized in the early 1990s, but miRNAs were not recognized as a distinct class of biologic regulators with conserved functions until the early 2000s. Since then, miRNA research has revealed multiple roles in negative regulation (transcript degradation and sequestering, translational suppression) and possible involvement in positive regulation (transcriptional and translational activation). By affecting gene regulation, miRNAs are likely to be involved in most biologic processes. Different sets of expressed miRNAs are found in different cell types and tissues.

Aberrant expression of miRNAs has been implicated in numerous disease states, and miRNA-based therapies are under investigation.

History

MicroRNAs were discovered in 1993 by Victor Ambros, Rosalind Lee and Rhonda Feinbaum during a study of the gene *lin-14* in *C. elegans* development. They found that LIN-14 protein abundance was regulated by a short RNA product encoded by the *lin-4* gene. A 61 nucleotide precursor from *lin-4* gene matured to a 22 nucleotide RNA containing sequences partially complementary to multiple sequences in the 3' UTR of the *lin-14* mRNA. This complementarity was sufficient and necessary to inhibit the translation of *lin-14* mRNA into LIN-14 protein. Retrospectively, the *lin-4* small RNA was the first microRNA to be identified, though at the time, it was thought to be a nematode idiosyncrasy. Only in 2000 was a second RNA characterized: let-7, which repressed *lin-41*, *lin-14*, *lin-28*, *lin-42*, and *daf-12* expression during developmental stage transitions in *C. elegans*. let-7 was soon found to be conserved in many species, indicating the existence of a wider phenomenon.

Nomenclature

Under a standard nomenclature system, names are assigned to experimentally confirmed miRNAs before publication of their discovery. The prefix "mir" is followed by a dash and a number, the latter often indicating order of naming. For example, mir-123 was named and likely discovered prior to mir-456. The uncapitalized "mir-" refers to the pre-miRNA, while a capitalized "miR-" refers to the mature form. miRNAs with nearly identical sequences bar one or two nucleotides are annotated with an additional lower case letter. For example, miR-123a would be closely related to miR-123b. Pre-miRNAs that lead to 100% identical mature miRNAs but that are located at different places in the genome are indicated with an additional dash-number suffix.

For example, the pre-miRNAs hsa-mir-194-1 and hsa-mir-194-2 lead to an identical mature miRNA (hsa-miR-194) but are located in different regions of the genome. Species of origin is designated with a three-letter prefix, e.g., hsa-miR-123 would be from human (*Homo sapiens*) and oar-miR-123 would be a sheep (*Ovis aries*) miRNA. Other common prefixes include 'v' for viral (miRNA encoded by a viral genome) and 'd' for *Drosophila* miRNA (a fruit fly commonly studied in genetic research). When two mature microRNAs originate from opposite arms of the same pre-miRNA, they are denoted with a -3p or -5p suffix. (In the past, this distinction was also made with 's' (sense) and 'as' (antisense)). When relative expression levels are known, an asterisk following the name indicates an miRNA expressed at low levels relative to the miRNA in the opposite arm of a hairpin. For example, miR-123 and miR-123* would share a pre-miRNA hairpin, but more miR-123 would be found in the cell.

Biogenesis

Most microRNA genes are found in intergenic regions or in anti-sense orientation to genes and contain their own miRNA gene promoter and regulatory units. As much as 40% of miRNA genes may lie in the introns of protein and non-protein coding genes or even in exons.

These are usually, though not exclusively, found in a sense orientation. and thus usually are regulated together with their host genes. Other miRNA genes showing a common promoter include the

42-48% of all miRNAs originating from polycistronic units containing 2-7 discrete loops from which mature miRNAs are processed, although this does not necessarily mean the mature miRNAs of a family will be homologous in structure and function. The promoters mentioned have been shown to have some similarities in their motifs to promoters of other genes transcribed by RNA polymerase II such as protein coding genes. The DNA template is not the final word on mature miRNA production: 6% of human miRNAs show RNA editing, the site-specific modification of RNA sequences to yield products different from those encoded by their DNA. This increases the diversity and scope of miRNA action beyond that implicated from the genome alone.

Transcription

miRNA genes are usually transcribed by RNA polymerase II (Pol II). The polymerase often binds to a promoter found near the DNA sequence encoding what will become the hairpin loop of the pre-miRNA. The resulting transcript is capped with a specially-modified nucleotide at the 5' end, polyadenylated with multiple adenosines (a poly(A) tail), and spliced. The product, called a primary miRNA (pri-miRNA), may be hundreds or thousands of nucleotides in length and contain one or more miRNA stem loops. When a stem loop precursor is found in the 3' UTR, a transcript may serve as a pri-miRNA and a mRNA. RNA polymerase III (Pol III) transcribes some miRNAs, especially those with upstream Alu sequences, transfer RNAs (tRNAs), and mammalian wide interspersed repeat (MWIR) promoter units.

Nuclear Processing

A single pri-miRNA may contain from one to six miRNA precursors. These hairpin loop structures are composed of about 70 nucleotides each. Each hairpin is flanked by sequences necessary for efficient processing. The double-stranded RNA structure of the hairpins in a pri-miRNA is recognized by a nuclear protein known as DiGeorge Syndrome Critical Region 8 (DGCR8 or "Pasha" in invertebrates), named for its association with DiGeorge Syndrome. DGCR8 associates with the enzyme Drosha, a protein that cuts RNA, to form the "Microprocessor" complex. In this complex, DGCR8 orients the catalytic RNase III domain of Drosha to liberate hairpins from pri-miRNAs by cleaving RNA about eleven nucleotides from the hairpin base (two helical RNA turns into the stem). The resulting hairpin, known as a pre-miRNA (precursor-miRNA), has a two-nucleotide overhang at its 3' end; it has 3' hydroxyl and 5' phosphate groups.

Pre-miRNAs that are spliced directly out of introns, bypassing the Microprocessor complex, are known as "mirtrons." Originally thought to exist only in *Drosophila* and *C. elegans*, mirtrons have now been found in mammals.

Perhaps as many as 16% of pri-miRNAs may be altered through nuclear RNA editing. Most commonly, enzymes known as adenosine deaminases acting on RNA (ADARs) catalyze adenosine to inosine (A to I) transitions. RNA editing can halt nuclear processing (for example, of pri-miR-142, leading to degradation by the ribonuclease Tudor-SN) and alter downstream processes including cytoplasmic miRNA processing and target specificity (e.g., by changing the seed region of miR-376 in the central nervous system).

Nuclear Export

pre-miRNA hairpins are exported from the nucleus in a process involving the nucleocytoplasmic shuttle Exportin-5. This protein, a member of the *karyopherin* family, recognizes a two-nucleotide overhang left by the RNase III enzyme Drosha at the 3' end of the pre-miRNA hairpin. Exportin-5-mediated transport to the cytoplasm is energy-dependent, using GTP bound to the Ran protein.

Cytoplasmic Processing

In the cytoplasm, the pre-miRNA hairpin is cleaved by the RNase III enzyme Dicer. This endoribonuclease interacts with the 3' end of the hairpin and cuts away the loop joining the 3' and 5' arms, yielding an imperfect miRNA:miRNA* duplex about 22 nucleotides in length. Overall hairpin length and loop size influence the efficiency of Dicer processing, and the imperfect nature of the miRNA:miRNA* pairing also affects cleavage. Although either strand of the duplex may potentially act as a functional miRNA, only one strand is usually incorporated into the RNA-induced silencing complex (RISC) where the miRNA and its mRNA target interact.

Biogenesis in Plants

miRNA biogenesis in plants differs from metazoan biogenesis mainly in the steps of nuclear processing and export. Instead of being cleaved by two different enzymes, once inside and once outside the nucleus, both cleavages of the plant miRNA is performed by a Dicer homolog, called Dicer-like1 (DL1). DL1 is only expressed in the nucleus of plant cells, which indicates that both reactions take place inside the nucleus. Before plant miRNA:miRNA* duplexes are transported out of the nucleus its 3' overhangs are methylated by a RNA methyltransferaseprotein called Hua-Enhancer1 (HEN1). The duplex is then transported out of the nucleus to the cytoplasm by a protein called Hasty (HST), an Exportin 5 homolog, where they disassemble and the mature miRNA is incorporated into the RISC.

The RNA-induced Silencing Complex

The mature miRNA is part of an active RNA-induced silencing complex (RISC) containing Dicer and many associated proteins. RISC is also known as a microRNA ribonucleoprotein complex (miRNP); RISC with incorporated miRNA is sometimes referred to as "miRISC."

Dicer processing of the pre-miRNA is thought to be coupled with unwinding of the duplex. Generally, only one strand is incorporated into the miRISC, selected on the basis of its thermodynamic instability and weaker base-pairing relative to the other strand. The position of the stem-loop may also influence strand choice. The other strand, called the passenger strand due to its lower levels in the steady state, is denoted with an asterisk (*) and is normally degraded. In some cases, both strands of the duplex are viable and become functional miRNA that target different mRNA populations.

Members of the argonaute (Ago) protein family are central to RISC function. Argonautes are needed for miRNA-induced silencing and contain two conserved RNA binding domains: a PAZ domain that can bind the single stranded 3' end of the mature miRNA and a PIWI domain that structurally resembles ribonuclease-H and functions to interact with the 5' end of the guide strand. They bind the mature miRNA and orient it for interaction with a target mRNA. Some argonautes, for example human Ago2, cleave target transcripts directly; argonautes may also recruit additional proteins to achieve translational repression. The human genome encodes eight argonaute proteins divided by sequence similarities into two families: AGO (with four members present in all mammalian cells and called E1F2C/hAgo in humans), and PIWI (found in the germ line and hematopoietic stem cells).

Additional RISC components include TRBP [human immunodeficiency virus (HIV) transactivating response RNA (TAR) binding protein], PACT (protein activator of the interferon induced protein kinase (PACT), the SMN complex, fragile X mental retardation protein (FMRP), and Tudor staphylococcal nuclease-domain-containing protein (Tudor-SN).

Mode of Silencing

Gene silencing may occur either via mRNA degradation or preventing mRNA from being translated. It has been demonstrated that if there is complete complementation between the miRNA and target mRNA sequence, Ago2 can cleave the mRNA and lead to direct mRNA degradation. Yet, if there isn't complete complementation the silencing is achieved by preventing translation.

miRNA Turnover

Turnover of mature miRNA is needed for rapid changes in miRNA expression profiles. During miRNA maturation in the cytoplasm, uptake by the Argonaute protein is thought to stabilize the guide strand, while the opposite (* or "passenger") strand is preferentially destroyed. In what has been called a "Use it or lose it" strategy, Argonaute may preferentially retain miRNAs with many targets over miRNAs with few or no targets, leading to degradation of the non-targeting molecules.

Decay of mature miRNAs in animals is mediated by the 5´-to-3´ exoribonuclease XRN2, also known as Rat1p. In plants, SDN (small RNA degrading nuclease) family members degrade miRNAs in the opposite (3'-to-5') direction. Similar enzymes are encoded in animal genomes, but their roles have not yet been described.

Several miRNA modifications affect miRNA stability. As indicated by work in the model organism *Arabidopsis thaliana* (thale cress), mature plant miRNAs appear to be stabilized by the addition of methyl moieties at the 3' end. The 2'-O-conjugated methyl groups block the addition of uracil (U) residues by uridyltransferase enzymes, a modification that may be associated with miRNA degradation. However, uridylation may also protect some miRNAs; the consequences of this modification are incompletely understood. Uridylation of some animal miRNAs has also been reported. Both plant and animal miRNAs may be altered by addition of adenine (A) residues to the 3' end of the miRNA. An extra A added to the end of mammalian miR-122, a liver-enriched miRNA important in Hepatitis C, stabilizes the molecule, and plant miRNAs ending with an adenine residue have slower decay rates.

Cellular Functions

The function of miRNAs appears to be in gene regulation. For that purpose, a miRNA is complementary to a part of one or more messenger RNAs (mRNAs). Animal miRNAs are usually complementary to a site in the 3' UTR whereas plant miRNAs are usually complementary to coding regions of mRNAs. Perfect or near perfect base pairing with the target RNA promotes cleavage of the RNA. This is the primary mode of plant microRNAs. In animals, microRNAs more often only partially base pair and inhibit protein translation of the target mRNA (this exists in plants as well but is less common). MicroRNAs that are partially complementary to a target can also speed up deadenylation, causing mRNAs to be degraded sooner. For partially complementary microRNAs to recognise their targets, nucleotides 2–7 of the miRNA (its 'seed region') still have to be perfectly complementary. miRNAs occasionally also cause histone modification and DNA methylation of promoter sites, which affects the expression of target genes.

Animal microRNAs target in particular developmental genes. In contrast, genes involved in functions common to all cells, such as gene expression, have very few microRNA target sites and seem to be under selection to avoid targeting by microRNAs.

dsRNA can also activate gene expression, a mechanism that has been termed "small RNA-induced gene activation" or RNAa. dsRNAs targeting gene promoters can induce potent transcriptional activation of associated genes. This was demonstrated in human cells using synthetic dsRNAs termed small activating RNAs (saRNAs), but has also been demonstrated for endogenous microRNA.

Evolution

MicroRNAs are significant phylogenetic markers because of their astonishingly low rate of evolution. Their origin may have permitted the development of morphological innovation, and by making gene expression more specific and 'fine-tunable', permitted the genesis of complex organs and perhaps, ultimately, complex life. Indeed, rapid bursts of morphological innovation are generally associated with a high rate of microRNA accumulation.

MicroRNAs originate predominantly by the random formation of hairpins in "non-coding" sections of DNA (i.e. introns or intergene regions), but also by the duplication and modification of existing microRNAs. The rate of evolution (i.e. nucleotide substitution) in recently-originated microRNAs is comparable to that elsewhere in the non-coding DNA, implying evolution by neutral drift; however, older microRNAs have a much lower rate of change (often less than one substitution per hundred million years), suggesting that once a microRNA gains a function it undergoes extreme purifying selection. At this point, a microRNA is rarely lost from an animal's genome, although microRNAs which are more recently derived (and thus presumably non-functional) are frequently lost. This makes them a valuable phylogenetic marker, and they are being looked upon as a possible solution to such outstanding phylogenetic problems as the relationships of arthropods.

MicroRNAs feature in the genomes of most eukaryotic organisms, from the brown algae to the metazoa. Across all species, in excess of 5000 had been identified by March 2010. Whilst short RNA sequences (50 – hundreds of base pairs) of a broadly comparable function occur in bacteria, bacteria lack true microRNAs.

Experimental Detection and Manipulation of miRNA

MicroRNA expression can be quantified in a two-step polymerase chain reaction process of modified RT-PCR followed by quantitative real-time PCR. Variations of this method achieve absolute or relative quantification. miRNAs can also be hybridized to microarrays, slides or chips with probes to hundreds or thousands of miRNA targets, so that relative levels of miRNAs can be determined in different samples. MicroRNAs can be both discovered and profiled by high-throughput sequencing methods. The activity of an miRNA can be experimentally inhibited using a locked nucleic acid (LNA) oligo, a Morpholino oligo or a 2'-O-methyl RNA oligo. MicroRNA maturation can be inhibited at several points by steric-blocking oligos. The miRNA target site of an mRNA transcript can also be blocked by a steric-blocking oligo. Additionally, a specific miRNA can be silenced by a complementary antagomir. For the "in situ" detection of miRNA, the use of LNA is currently the only efficient method. The locked conformation of LNA results in enhanced hybridization properties and increases sensitivity and selectivity, making it ideal for detection of short miRNA.

miRNA and Disease

Just as miRNA is involved in the normal functioning of eukaryotic cells, so has dysregulation of miRNA been associated with disease. A manually curated, publicly available database miR2Disease documents known relationships between miRNA dysregulation and human disease.

miRNA and Cancer

Several miRNAs have been found to have links with some types of cancer. A study of mice altered to produce excess c-Myc — a protein with mutated forms implicated in several cancers — shows that miRNA has an effect on the development of cancer. Mice that were engineered to produce a surplus of types of miRNA found in lymphoma cells developed the disease within 50 days and died two weeks later. In contrast, mice without the surplus miRNA lived over 100 days. Leukemia can be caused by the insertion of a viral genome next to the 17-92 array of microRNAs leading to increased expression of this microRNA.

Another study found that two types of miRNA inhibit the E2F1 protein, which regulates cell proliferation. miRNA appears to bind to messenger RNA before it can be translated to proteins that switch genes on and off. By measuring activity among 217 genes encoding miRNA, patterns of gene activity that can distinguish types of cancers can be discerned. miRNA signatures may enable classification of cancer. This will allow doctors to determine the original tissue type which spawned a cancer and to be able to target a treatment course based on the original tissue type. miRNA profiling has already been able to determine whether patients with chronic lymphocytic leukemia had slow growing or aggressive forms of the cancer.

Transgenic mice that over-express or lack specific miRNAs have provided insight into the role of small RNAs in various malignancies.

A novel miRNA-profiling based screening assay for the detection of early-stage colorectal cancer has been developed and is currently in clinical trials. Early results showed that blood plasma samples collected from patients with early, resectable (Stage II) colorectal cancer could be distinguished from those of sex-and age-matched healthy volunteers. Sufficient selectivity and specificity could be achieved using small (less than 1 mL) samples of blood. The test has potential to be a cost-effective, non-invasive way to identify at-risk patients who should undergo colonoscopy.

miRNA and Heart Disease

The global role of miRNA function in the heart has been addressed by conditionally inhibiting miRNA maturation in the murine heart, and has revealed that miRNAs play an essential role during its development. miRNA expression profiling studies demonstrate that expression levels of specific miRNAs change in diseased human hearts, pointing to their involvement in cardiomyopathies. Furthermore, studies on specific miRNAs in animal models have identified distinct roles for miRNAs both during heart development and under pathological conditions, including the regulation of key factors important for cardiogenesis, the hypertrophic growth response, and cardiac conductance.

miRNA and the Nervous System

miRNAs appear to regulate the nervous system. Neural miRNAs are involved at various stages of synaptic development, including dendritogenesis (involving miR-132, miR-134 and miR-124), synapse formation and synapse maturation (where miR-134 and miR-138 are thought to be involved). Some studies find altered miRNA expression in schizophrenia.

miRNA and Non-coding RNAs

When the human genome project mapped its first chromosome in 1999, it was predicted the genome would contain over 100,000 protein coding genes. However, only around 20,000 were eventually identified (International Human Genome Sequencing Consortium, 2004). Since then, the advent of bioinformatics approaches combined with genome tiling studies examining the transcriptome, systematic sequencing of full length cDNA libraries, and experimental validation (including the creation of miRNA derived antisense oligonucleotides called antagomirs) have revealed that many transcripts are non protein-coding RNA, including several snoRNAs and miRNAs.

Other Decay Mechanisms

There are other ways in which messages can be degraded, including non-stop decay, silencing by Piwi-interacting RNA (piRNA), and surely other means.

6

Genetic Material: Properties and Replication

DNA

Deoxyribonucleic acid (DNA) is a nucleic acid that contains the genetic instructions used in the development and functioning of all known living organisms with the exception of some viruses. The main role of DNA molecules is the long-term storage of information. DNA is often compared to a set of blueprints, like a recipe or a code, since it contains the instructions needed to construct other components of cells, such as proteins and RNA molecules. The DNA segments that carry this genetic information are called genes, but other DNA sequences have structural purposes, or are involved in regulating the use of this genetic information.

DNA consists of two long polymers of simple units called nucleotides, with backbones made of sugars and phosphate groups joined by ester bonds. These two strands run in opposite directions to each other and are therefore anti-parallel. Attached to each sugar is one of four types of molecules called bases. It is the sequence of these four bases along the backbone that encodes information. This information is read using the genetic code, which specifies the sequence of the amino acids within proteins. The code is read by copying stretches of DNA into the related nucleic acid RNA, in a process called transcription.

Within cells, DNA is organized into long structures called chromosomes. These chromosomes are duplicated before cells divide, in a process called DNA replication. Eukaryotic organisms (animals, plants, fungi, and protists) store most of their DNA inside the cell nucleus and some of their DNA in organelles, such as mitochondria or chloroplasts. In contrast, prokaryotes (bacteria and archaea) store their DNA only in the cytoplasm. Within the chromosomes, chromatin proteins such as histones compact and organize DNA. These compact structures guide the interactions between DNA and other proteins, helping control which parts of the DNA are transcribed.

Properties

DNA is a long polymer made from repeating units called nucleotides. The DNA chain is 22 to 26 Angströms wide (2.2 to 2.6 nanometres), and one nucleotide unit is 3.3 Å (0.33 nm) long. Although each individual repeating unit is very small, DNA polymers can be very large molecules containing millions of nucleotides. For instance, the largest human chromosome, chromosome number 1, is approximately 220 million base pairs long. In living organisms, DNA does not usually exist as a single molecule, but instead as a pair of molecules that are held tightly together. These two long strands entwine like vines, in the shape of a double helix. The nucleotide repeats contain both the segment of the backbone of the molecule, which holds the chain together, and a base, which interacts with the other DNA strand in the helix. A base linked to a sugar is called a nucleoside and a base linked to a sugar and one or more phosphate groups is called a nucleotide. If multiple nucleotides are linked together, as in DNA, this polymer is called a polynucleotide.

The backbone of the DNA strand is made from alternating phosphate and sugar residues. The sugar in DNA is 2-deoxyribose, which is a pentose (five-carbon) sugar. The sugars are joined together by phosphate groups that form phosphodiester bonds between the third and fifth carbon atoms of adjacent sugar rings. These asymmetric bonds mean a strand of DNA has a direction. In a double helix the direction of the nucleotides in one strand is opposite to their direction in the other strand: the strands are *antiparallel*. The asymmetric ends of DNA strands are called the 52 (*five prime*) and 32 (*three prime*) ends, with the 5' end having a terminal phosphate group and the 3' end a terminal hydroxyl group. One major difference between DNA and RNA is the sugar, with the 2-deoxyribose in DNA being replaced by the alternative pentose sugar ribose in RNA.

The DNA double helix is stabilized by hydrogen bonds between the bases attached to the two strands. The four bases found in DNA are adenine (abbreviated A), cytosine (C), guanine (G) and thymine

(T). These four bases are attached to the sugar/phosphate to form the complete nucleotide, as shown for adenosine monophosphate.

These bases are classified into two types; adenine and guanine are fused five- and six-membered heterocyclic compounds called purines, while cytosine and thymine are six-membered rings called pyrimidines. A fifth pyrimidine base, called uracil (U), usually takes the place of thymine in RNA and differs from thymine by lacking a methyl group on its ring. Uracil is not usually found in DNA, occurring only as a breakdown product of cytosine. In addition to RNA and DNA, a large number of artificial nucleic acid analogues have also been created to study the proprieties of nucleic acids, or for use in biotechnology.

Grooves

Twin helical strands form the DNA backbone. Another double helix may be found by tracing the spaces, or grooves, between the strands. These voids are adjacent to the base pairs and may provide a binding site. As the strands are not directly opposite each other, the grooves are unequally sized. One groove, the major groove, is 22 wide and the other, the minor groove, is 12 wide. The narrowness of the minor groove means that the edges of the bases are more accessible in the major groove. As a result, proteins like transcription factors that can bind to specific sequences in double-stranded DNA usually make contacts to the sides of the bases exposed in the major groove. This situation varies in unusual conformations of DNA within the cell, but the major and minor grooves are always named to reflect the differences in size that would be seen if the DNA is twisted back into the ordinary B form.

Base Pairing

Each type of base on one strand forms a bond with just one type of base on the other strand. This is called complementary base pairing. Here, purines form hydrogen bonds to pyrimidines, with A bonding only to T, and C bonding only to G. This arrangement of two nucleotides binding together across the double helix is called a base pair.

As hydrogen bonds are not covalent, they can be broken and rejoined relatively easily. The two strands of DNA in a double helix can therefore be pulled apart like a zipper, either by a mechanical force or high temperature. As a result of this complementarity, all the information in the double-stranded sequence of a DNA helix is duplicated on each strand, which is vital in DNA replication. Indeed, this reversible and specific interaction between complementary base pairs is critical for all the functions of DNA in living organisms.

The two types of base pairs form different numbers of hydrogen bonds, AT forming two hydrogen bonds, and GC forming three hydrogen bonds. DNA with high GC-content is more stable than DNA with low GC-content, but contrary to popular belief, this is not due to the extra hydrogen bond of a GC base pair but rather the contribution of stacking interactions (hydrogen bonding merely provides specificity of the pairing, not stability). As a result, it is both the percentage of GC base pairs and the overall length of a DNA double helix that determine the strength of the association between the two strands of DNA.

Long DNA helices with a high GC content have stronger-interacting strands, while short helices with high AT content have weaker-interacting strands. In biology, parts of the DNA double helix that need to separate easily, such as the TATAAT Pribnow box in some promoters, tend to have a high AT content, making the strands easier to pull apart. In the laboratory, the strength of this interaction can be measured by finding the temperature required to break the hydrogen bonds, their melting temperature (also called T_m value). When all the base pairs in a DNA double helix melt, the strands separate and exist in solution as two entirely independent molecules. These single-stranded DNA molecules (*ssDNA*) have no single common shape, but some conformations are more stable than others.

Sense and Antisense

A DNA sequence is called "sense" if its sequence is the same as that of a messenger RNA copy that is translated into protein. The sequence on the opposite strand is called the "antisense" sequence. Both sense and antisense sequences can exist on different parts of the same strand of DNA (i.e. both strands contain both sense and antisense sequences). In both prokaryotes and eukaryotes, antisense RNA sequences are produced, but the functions of these RNAs are not entirely clear. One proposal is that antisense RNAs are involved in regulating gene expression through RNA-RNA base pairing.

A few DNA sequences in prokaryotes and eukaryotes, and more in plasmids and viruses, blur the distinction between sense and antisense strands by having overlapping genes. In these cases, some DNA sequences do double duty, encoding one protein when read along one strand, and a second protein when read in the opposite direction along the other strand. In bacteria, this overlap may be involved in the regulation of gene transcription, while in viruses, overlapping genes increase the amount of information that can be encoded within the small viral genome.

Supercoiling

DNA can be twisted like a rope in a process called DNA supercoiling. With DNA in its "relaxed" state, a strand usually circles the axis of the double helix once every 10.4 base pairs, but if the DNA is twisted the strands become more tightly or more loosely wound. If the DNA is twisted in the direction of the helix, this is positive supercoiling, and the bases are held more tightly together. If they are twisted in the opposite direction, this is negative supercoiling, and the bases come apart more easily. In nature, most DNA has slight negative supercoiling that is introduced by enzymes called topoisomerases. These enzymes are also needed to relieve the twisting stresses introduced into DNA strands during processes such as transcription and DNA replication.

Alternate DNA Structures

DNA exists in many possible conformations that include A-DNA, B-DNA, and Z-DNA forms, although, only B-DNA and Z-DNA have been directly observed in functional organisms. The conformation that DNA adopts depends on the hydration level, DNA sequence, the amount and direction of supercoiling, chemical modifications of the bases, the type and concentration of metal ions, as well as the presence of polyamines in solution.

The first published reports of A-DNA X-ray diffraction patterns— and also B-DNA used analyses based on Patterson transforms that provided only a limited amount of structural information for oriented fibres of DNA. An alternate analysis was then proposed by Wilkins *et al.*, in 1953, for the *in vivo* B-DNA X-ray diffraction/scattering patterns of highly hydrated DNA fibres in terms of squares of Bessel functions. In the same journal, James D. Watson and Francis Crick presented their molecular modeling analysis of the DNA X-ray diffraction patterns to suggest that the structure was a double-helix.

Although the 'B-DNA form' is most common under the conditions found in cells, it is not a well-defined conformation but a family of related DNA conformations that occur at the high hydration levels present in living cells. Their corresponding X-ray diffraction and scattering patterns are characteristic of molecular paracrystals with a significant degree of disorder.

Compared to B-DNA, the A-DNA form is a wider right-handed spiral, with a shallow, wide minor groove and a narrower, deeper major groove. The A form occurs under non-physiological conditions in partially dehydrated samples of DNA, while in the cell it may be produced in hybrid pairings of DNA and RNA strands, as well as in enzyme-DNA complexes. Segments of DNA where the bases have been chemically modified by methylation may undergo a larger change in conformation and adopt the Z form. Here, the strands turn about the helical axis in a left-handed spiral, the opposite of the more common B form. These unusual structures can be recognized by specific Z-DNA binding proteins and may be involved in the regulation of transcription.

Quadruplex Structures

At the ends of the linear chromosomes are specialized regions of DNA called telomeres. The main function of these regions is to allow the cell to replicate chromosome ends using the enzyme telomerase, as the enzymes that normally replicate DNA cannot copy the extreme 32 ends of chromosomes. These specialized chromosome caps also help protect the DNA ends, and stop the DNA repair systems in the cell from treating them as damage to be corrected. In human cells, telomeres are usually lengths of single-stranded DNA containing several thousand repeats of a simple TTAGGG sequence.

These guanine-rich sequences may stabilize chromosome ends by forming structures of stacked sets of four-base units, rather than the usual base pairs found in other DNA molecules. Here, four guanine bases form a flat plate and these flat four-base units then stack on top of each other, to form a stable *G-quadruplex* structure. These structures are stabilized by hydrogen bonding between the edges of the bases and chelation of a metal ion in the centre of each four-base unit. Other structures can also be formed, with the central set of four bases coming from either a single strand folded around the bases, or several different parallel strands, each contributing one base to the central structure.

In addition to these stacked structures, telomeres also form large loop structures called telomere loops, or T-loops. Here, the single-stranded DNA curls around in a long circle stabilized by telomere-binding proteins. At the very end of the T-loop, the single-stranded telomere DNA is held onto a region of double-stranded DNA by the telomere strand disrupting the double-helical DNA and base pairing to one of the two strands. This triple-stranded structure is called a displacement loop or D-loop.

Branched DNA

In DNA fraying occurs when non-complementary regions exist at the end of an otherwise complementary double-strand of DNA. However, branched DNA can occur if a third strand of DNA is introduced and contains adjoining regions able to hybridize with the frayed regions of the pre-existing double-strand. Although the simplest example of branched DNA involves only three strands of DNA, complexes involving additional strands and multiple branches are also possible.

Vibration

DNA may carry out low-frequency collective motion as observed by the Raman spectroscopy and analyzed with a quasi-continuum model.

Chemical Modifications

Base Modifications

The expression of genes is influenced by how the DNA is packaged in chromosomes, in a structure called chromatin. Base modifications can be involved in packaging, with regions that have low or no gene expression usually containing high levels of methylation of cytosine bases. For example, cytosine methylation, produces 5-methylcytosine, which is important for X-chromosome inactivation. The average level of methylation varies between organisms - the worm *Caenorhabditis elegans* lacks cytosine methylation, while vertebrates have higher levels, with up to 1% of their DNA containing 5-methylcytosine. Despite the importance of 5-methylcytosine, it can deaminate to leave a thymine base, so methylated cytosines are particularly prone to mutations. Other base modifications include adenine methylation in bacteria, the presence of 5-hydroxymethylcytosine in the brain, and the glycosylation of uracil to produce the "J-base" in kinetoplastids.

Damage

DNA can be damaged by many sorts of mutagens, which change the DNA sequence. Mutagens include oxidizing agents, alkylating agents and also high-energy electromagnetic radiation such as ultraviolet light and X-rays. The type of DNA damage produced depends on the type of mutagen. For example, UV light can damage DNA by producing thymine dimers, which are cross-links between pyrimidine bases. On the other hand, oxidants such as free radicals or hydrogen peroxide produce multiple forms of damage, including base modifications, particularly of guanosine, and double-strand breaks. A typical human cell contains about 150,000 bases that have suffered oxidative damage. Of these oxidative lesions, the most dangerous are double-strand breaks, as these are difficult to repair and can produce point mutations, insertions and deletions from the DNA sequence, as well as chromosomal translocations.

Many mutagens fit into the space between two adjacent base pairs, this is called *intercalation*. Most intercalators are aromatic and planar molecules; examples include ethidium bromide, daunomycin, and doxorubicin. In order for an intercalator to fit between base pairs, the bases must separate, distorting the DNA strands by unwinding of the double helix. This inhibits both transcription and DNA replication, causing toxicity and mutations. As a result, DNA intercalators are often carcinogens, and benzo[*a*]pyrene diol epoxide, acridines, aflatoxin and ethidium bromide are well-known examples. Nevertheless, due to their ability to inhibit DNA transcription and replication, other similar toxins are also used in chemotherapy to inhibit rapidly growing cancer cells.

Biological Functions

DNA usually occurs as linear chromosomes in eukaryotes, and circular chromosomes in prokaryotes. The set of chromosomes in a cell makes up its genome; the human genome has approximately 3 billion base pairs of DNA arranged into 46 chromosomes. The information carried by DNA is held in the sequence of pieces of DNA called genes. Transmission of genetic information in genes is achieved via complementary base pairing.

For example, in transcription, when a cell uses the information in a gene, the DNA sequence is copied into a complementary RNA sequence through the attraction between the DNA and the correct RNA nucleotides. Usually, this RNA copy is then used to make a matching protein sequence in a process called translation, which depends on the same interaction between RNA nucleotides. In alternative fashion, a cell may simply copy its genetic information in a process called DNA replication. The details of these functions are covered in other articles; here we focus on the interactions between DNA and other molecules that mediate the function of the genome.

Genes and Genomes

Genomic DNA is tightly and orderly packed in the process called DNA condensation to fit the small available volumes of the cell. In eukaryotes, DNA is located in the cell nucleus, as well as small amounts in mitochondria and chloroplasts. In prokaryotes, the DNA is held within an irregularly shaped body in the cytoplasm called the nucleoid. The genetic information in a genome is held within genes, and the complete set of this information in an organism is called its genotype. A gene is a unit of heredity and is a region of DNA that influences a particular characteristic in an organism. Genes contain an open reading frame that can be transcribed, as well as regulatory sequences such as promoters and enhancers, which control the transcription of the open reading frame.

In many species, only a small fraction of the total sequence of the genome encodes protein. For example, only about 1.5% of the human genome consists of protein-coding exons, with over 50% of human DNA consisting of non-coding repetitive sequences. The reasons for the presence of so much non-coding DNA in eukaryotic genomes and the extraordinary differences in genome size, or *C-value*, among species represent a long-standing puzzle known as the "C-value enigma". However, DNA sequences that do not code protein may still encode functional non-coding RNA molecules, which are involved in the regulation of gene expression.

Some non-coding DNA sequences play structural roles in chromosomes. Telomeres and centromeres typically contain few genes, but are important for the function and stability of chromosomes. An abundant form of non-coding DNA in humans are pseudogenes, which are copies of genes that have been disabled by mutation. These sequences are usually just molecular fossils, although they can occasionally serve as raw genetic material for the creation of new genes through the process of gene duplication and divergence.

Transcription and Translation

A gene is a sequence of DNA that contains genetic information and can influence the phenotype of an organism. Within a gene, the sequence of bases along a DNA strand defines a messenger RNA sequence, which then defines one or more protein sequences. The relationship between the nucleotide sequences of genes and the amino-acid sequences of proteins is determined by the rules of translation, known collectively as the genetic code. The genetic code consists of three-letter 'words' called *codons* formed from a sequence of three nucleotides (e.g. ACT, CAG, TTT). In transcription, the codons of a gene are copied into messenger RNA by RNA polymerase. This RNA copy is then decoded by a ribosome that reads the RNA sequence by base-pairing the messenger RNA to transfer RNA, which carries amino acids. Since there are 4 bases in 3-letter combinations, there are 64 possible codons (4 combinations). These encode the twenty standard amino acids, giving most amino acids more than one possible codon. There are also three 'stop' or 'nonsense' codons signifying the end of the coding region; these are the TAA, TGA and TAG codons.

Replication

Cell division is essential for an organism to grow, but, when a cell divides, it must replicate the DNA in its genome so that the two daughter cells have the same genetic information as their parent. The double-stranded structure of DNA provides a simple mechanism for DNA replication. Here, the two strands are separated and then each strand's complementary DNA sequence is recreated by an enzyme called DNA polymerase. This enzyme makes the complementary strand by finding the correct base through complementary base pairing, and bonding it onto the original strand. As DNA polymerases can only extend a DNA strand in a 52 to 32 direction, different mechanisms are used to copy the antiparallel strands of the double helix. In this way, the base on the old strand dictates which base appears on the new strand, and the cell ends up with a perfect copy of its DNA.

Interactions with Proteins

All the functions of DNA depend on interactions with proteins. These protein interactions can be non-specific, or the protein can bind specifically to a single DNA sequence. Enzymes can also bind to DNA and of these, the polymerases that copy the DNA base sequence in transcription and DNA replication are particularly important.

DNA-binding Proteins

Structural proteins that bind DNA are well-understood examples of non-specific DNA-protein interactions. Within chromosomes, DNA is held in complexes with structural proteins. These proteins organize the DNA into a compact structure called chromatin. In eukaryotes this structure involves DNA binding to a complex of small basic proteins called histones, while in prokaryotes multiple types of proteins are involved. The histones form a disk-shaped complex called a nucleosome, which contains two complete turns of double-stranded DNA wrapped around its surface.

These non-specific interactions are formed through basic residues in the histones making ionic bonds to the acidic sugar-phosphate backbone of the DNA, and are therefore largely independent of the base sequence. Chemical modifications of these basic amino acid residues include methylation, phosphorylation and acetylation. These chemical changes alter the strength of the interaction between the DNA and the histones, making the DNA more or less accessible to transcription factors and changing the rate of transcription. Other non-specific DNA-binding proteins in chromatin include the high-mobility group proteins, which bind to bent or distorted DNA. These proteins are important in bending arrays of nucleosomes and arranging them into the larger structures that make up chromosomes.

A distinct group of DNA-binding proteins are the DNA-binding proteins that specifically bind single-stranded DNA. In humans, replication protein A is the best-understood member of this family and is used in processes where the double helix is separated, including DNA replication, recombination and DNA repair. These binding proteins seem to stabilize single-stranded DNA and protect it from forming stem-loops or being degraded by nucleases.

In contrast, other proteins have evolved to bind to particular DNA sequences. The most intensively studied of these are the various transcription factors, which are proteins that regulate transcription. Each transcription factor binds to one particular set of DNA sequences and activates or inhibits the transcription of genes that have these sequences close to their promoters. The transcription factors do this in two ways. Firstly, they can bind the RNA polymerase responsible for transcription, either directly or through other mediator proteins; this locates the polymerase at the promoter and allows it to begin transcription. Alternatively, transcription factors can bind enzymes that modify the histones at the promoter; this will change the accessibility of the DNA template to the polymerase.

As these DNA targets can occur throughout an organism's genome, changes in the activity of one type of transcription factor can affect thousands of genes. Consequently, these proteins are often the targets of the signal transduction processes that control responses to environmental changes or cellular differentiation and development. The specificity of these transcription factors' interactions with DNA come from the proteins making multiple contacts to the edges of the DNA bases, allowing them to "read" the DNA sequence. Most of these base-interactions are made in the major groove, where the bases are most accessible.

DNA-modifying Enzymes

Nucleases and Ligases

Nucleases are enzymes that cut DNA strands by catalyzing the hydrolysis of the phosphodiester bonds. Nucleases that hydrolyse nucleotides from the ends of DNA strands are called exonucleases, while endonucleases cut within strands. The most frequently used nucleases in molecular biology are the restriction endonucleases, which cut DNA at specific sequences. For instance, the EcoRV enzyme shown to the left recognizes the 6-base sequence 52 -GAT | ATC-32 and makes a cut at the vertical line. In nature, these enzymes protect bacteria against phage infection by digesting the phage DNA when it enters the bacterial cell, acting as part of the restriction modification system. In technology, these sequence-specific nucleases are used in molecular cloning and DNA fingerprinting.

Enzymes called DNA ligases can rejoin cut or broken DNA strands. Ligases are particularly important in lagging strand DNA replication, as they join together the short segments of DNA produced at the replication fork into a complete copy of the DNA template. They are also used in DNA repair and genetic recombination.

Topoisomerases and Helicases

Topoisomerases are enzymes with both nuclease and ligase activity. These proteins change the amount of supercoiling in DNA. Some of these enzymes work by cutting the DNA helix and allowing one section to rotate, thereby reducing its level of supercoiling; the enzyme then seals the DNA break. Other types of these enzymes are capable of cutting one DNA helix and then passing a second strand of DNA through this break, before rejoining the helix. Topoisomerases are required for many processes involving DNA, such as DNA replication and transcription. Helicases are proteins that are a type of molecular motor. They use the chemical energy in nucleoside triphosphates, predominantly ATP, to break hydrogen bonds between bases and unwind the DNA double helix into single strands. These enzymes are essential for most processes where enzymes need to access the DNA bases.

Polymerases

Polymerases are enzymes that synthesize polynucleotide chains from nucleoside triphosphates. The sequence of their products are copies of existing polynucleotide chains - which are called *templates*. These enzymes function by adding nucleotides onto the 32 hydroxyl group of the previous nucleotide in a DNA strand. As a consequence, all polymerases work in a 52 to 32 direction. In the active site of these enzymes, the incoming nucleoside triphosphate base-pairs to the template: this allows polymerases to accurately synthesize the complementary strand of their template. Polymerases are classified according to the type of template that they use.

In DNA replication, a DNA-dependent DNA polymerase makes a copy of a DNA sequence. Accuracy is vital in this process, so many of these polymerases have a proofreading activity. Here, the polymerase recognizes the occasional mistakes in the synthesis reaction by the lack of base pairing between the mismatched nucleotides. If a mismatch is detected, a 32 to 52 exonuclease activity is activated and the incorrect base removed. In most organisms, DNA polymerases function in a large complex called the replisome that contains multiple accessory subunits, such as the DNA clamp or helicases.

RNA-dependent DNA polymerases are a specialized class of polymerases that copy the sequence of an RNA strand into DNA. They include reverse transcriptase, which is a viral enzyme involved in the infection of cells by retroviruses, and telomerase, which is required for the replication of telomeres. Telomerase is an unusual polymerase because it contains its own RNA template as part of its structure.

Transcription is carried out by a DNA-dependent RNA polymerase that copies the sequence of a DNA strand into RNA. To begin transcribing a gene, the RNA polymerase binds to a sequence of DNA called a promoter and separates the DNA strands. It then copies the gene sequence into a messenger RNA transcript until it reaches a region of DNA called the terminator, where it halts and detaches from the DNA. As with human DNA-dependent DNA polymerases, RNA polymerase II, the enzyme that transcribes most of the genes in the human genome, operates as part of a large protein complex with multiple regulatory and accessory subunits.

Genetic Recombination

A DNA helix usually does not interact with other segments of DNA, and in human cells the different chromosomes even occupy separate areas in the nucleus called "chromosome territories". This physical separation of different chromosomes is important for the ability of DNA to function as a stable repository for information, as one of the few times chromosomes interact is during chromosomal crossover when they recombine. Chromosomal crossover is when two DNA helices break, swap a section and then rejoin.

Recombination allows chromosomes to exchange genetic information and produces new combinations of genes, which increases the efficiency of natural selection and can be important in the rapid evolution of new proteins. Genetic recombination can also be involved in DNA repair, particularly in the cell's response to double-strand breaks. The most common form of chromosomal crossover is homologous recombination, where the two chromosomes involved share very similar sequences. Non-homologous recombination can be damaging to cells, as it can produce chromosomal translocations and genetic abnormalities. The recombination reaction is catalyzed by enzymes known as recombinases, such as RAD51.

The first step in recombination is a double-stranded break either caused by an endonuclease or damage to the DNA. A series of steps catalyzed in part by the recombinase then leads to joining of the two helices by at least one Holliday junction, in which a segment of a single strand in each helix is annealed to the complementary strand in the other helix. The Holliday junction is a tetrahedral junction structure that can be moved along the pair of chromosomes, swapping one strand for another. The recombination reaction is then halted by cleavage of the junction and re-ligation of the released DNA.

Evolution

DNA contains the genetic information that allows all modern living things to function, grow and reproduce. However, it is unclear how long in the 4-billion-year history of life DNA has performed this function, as it has been proposed that the earliest forms of life may have used RNA as their genetic material. RNA may have acted as the central part of early cell metabolism as it can both transmit genetic information and carry out catalysis as part of ribozymes. This ancient RNA world where nucleic acid would have been used for both catalysis and genetics may have influenced the evolution of the current genetic code based on four nucleotide bases. This would occur, since the number of different bases in such an organism is a trade-off between a small number of bases increasing replication accuracy and a large number of bases increasing the catalytic efficiency of ribozymes.

However, there is no direct evidence of ancient genetic systems, as recovery of DNA from most fossils is impossible. This is because DNA will survive in the environment for less than one million years and slowly degrades into short fragments in solution. Claims for older DNA have been made, most notably a report of the isolation of a viable bacterium from a salt crystal 250 million years old, but these claims are controversial.

Uses in Technology

Genetic Engineering

Methods have been developed to purify DNA from organisms, such as phenol-chloroform extraction and manipulate it in the laboratory, such as restriction digests and the polymerase chain reaction. Modern biology and biochemistry make intensive use of these techniques in recombinant DNA technology. Recombinant DNA is a man-made DNA sequence that has been assembled from other DNA sequences. They can be transformed into organisms in the form of plasmids or in the appropriate format, by using a viral vector. The genetically modified organisms produced can be used to produce products such as recombinant proteins, used in medical research, or be grown in agriculture.

Forensics

Forensic scientists can use DNA in blood, semen, skin, saliva or hair found at a crime scene to identify a matching DNA of an individual, such as a perpetrator. This process is formally termed DNA profiling, but may also be called "genetic fingerprinting". In DNA profiling, the lengths of variable sections of repetitive DNA, such as short tandem repeats and minisatellites, are compared between people. This method is usually an extremely reliable technique for identifying a matching DNA. However, identification can be complicated if the scene is contaminated with DNA from several people. DNA profiling was developed in 1984 by British geneticist Sir Alec Jeffreys, and first used in forensic science to convict Colin Pitchfork in the 1988 Enderby murders case.

People convicted of certain types of crimes may be required to provide a sample of DNA for a database. This has helped investigators solve old cases where only a DNA sample was obtained from the scene. DNA profiling can also be used to identify victims of mass casualty incidents. On the other hand, many convicted people have been released from prison on the basis of DNA techniques, which were not available when a crime had originally been committed.

DNA Profiling

DNA profiling (also called DNA testing, DNA typing, or genetic fingerprinting) is a technique employed by forensic scientists to assist in the identification of individuals on the basis of their respective DNA profiles. DNA profiles are encrypted sets of numbers that reflect a person's DNA make-up, which can also be used as the person's identifier. DNA profiling should not be confused with full genome sequencing. It is used in, for example, parental testing and rape investigation.

Although 99.9% of human DNA sequences are the same in every person, enough of the DNA is different to distinguish one individual from another. DNA profiling uses repetitive ("repeat") sequences that are highly variable, called variable number tandem repeats (VNTR). VNTRs loci are very similar between closely related humans, but so variable that unrelated individuals are extremely unlikely to have the same VNTRs. The DNA profiling technique was first reported in 1984 by Sir Alec Jeffreys at the University of Leicester in England, and is now the basis of several national DNA databases. Dr. Jeffreys's genetic fingerprinting was made commercially available in 1987, when a chemical company, ICI, started a blood-testing centre in England.

DNA Profiling Process

The process begins with a sample of an individual's DNA (typically called a "reference sample"). The most desirable method of collecting a reference sample is the use of a buccal swab, as this reduces the possibility of contamination. When this is not available (e.g. because a court order may be needed and not obtainable) other methods may need to be used to collect a sample of blood, saliva, semen, or other appropriate fluid or tissue from personal items (e.g. toothbrush, razor, etc.) or from stored samples (e.g. banked sperm or biopsy tissue). Samples obtained from blood relatives (biological relative) can provide an indication of an individual's profile, as could human remains which had been previously profiled.

A reference sample is then analyzed to create the individual's DNA profile using one of a number of techniques, discussed below. The DNA profile is then compared against another sample to determine whether there is a genetic match.

RFLP Analysis

The first methods for finding out genetics used for DNA profiling involved restriction enzyme digestion, followed by Southern blot analysis. Although polymorphisms can exist in the restriction enzyme cleavage sites, more commonly the enzymes and DNA probes were used to analyze VNTR loci. However, the Southern blot technique is laborious, and requires large amounts of undegraded sample DNA. Also, Karl Brown's original technique looked at many minisatellite loci at the same time, increasing the observed variability, but making it hard to discern individual alleles (and thereby precluding parental testing). These early techniques have been supplanted by PCR-based assays.

PCR Analysis

With the invention of the polymerase chain reaction (PCR) technique, DNA profiling took huge strides forward in both discriminating power and the ability to recover information from very small (or degraded) starting samples. PCR greatly amplifies the amounts of a specific region of DNA, using oligonucleotide primers and a thermostable DNA polymerase.

Early assays such as the HLA-DQ alpha reverse dot blot strips grew to be very popular due to their ease of use, and the speed with which a result could be obtained. However they were not as discriminating as RFLP. It was also difficult to determine a DNA profile for mixed samples, such as a vaginal swab from a sexual assault victim.

Fortunately, the PCR method is readily adaptable for analyzing VNTR loci. In the United States the FBI has standardized a set of 13 VNTR assays for DNA typing, and has organized the CODIS database for forensic identification in criminal cases. Similar assays and databases have been set up in other countries. Also, commercial kits are available that analyze single-nucleotide polymorphisms (SNPs). These kits use PCR to amplify specific regions with known variations and hybridize them to probes anchored on cards, which results in a coloured spot corresponding to the particular sequence variation.

STR Analysis

The method of DNA profiling used today is based on PCR and uses short tandem repeats (STR). This method uses highly polymorphic regions that have short repeated sequences of DNA (the most common is 4 bases repeated, but there are other lengths in use, including 3 and 5 bases). Because unrelated people almost certainly have different numbers of repeat units, STRs can be used to discriminate between unrelated individuals. These STR loci (locations on a chromosome) are targeted with sequence-specific primers and amplified using PCR. The DNA fragments that result are then separated and detected using electrophoresis. There are two common methods of separation and detection, capillary electrophoresis (CE) and gel electrophoresis.

Each STR is polymorphic, however, the number of alleles is small. Typically each STR allele will be shared by around 5 - 20% of individuals. The power of STR analysis comes from looking at multiple STR loci simultaneously. The pattern of alleles can identify an individual quite accurately. Thus STR analysis provides an excellent identification tool. The more STR regions that are tested in an individual the more discriminating the test becomes.

From country to country, different STR-based DNA-profiling systems are in use. In North America, systems which amplify the CODIS 13 core loci are almost universal, while in the UK the SGM+ system (which is compatible with The National DNA Database), is in use. Whichever system is used, many of the STR regions used are the same. These DNA-profiling systems are based on multiplex reactions, whereby many STR regions will be tested at the same time.

The true power of STR analysis is in its statistical power of discrimination. Because the 13 loci that are currently used for discrimination in CODIS are independently assorted (having a certain number of repeats at one locus doesn't change the likelihood of having any number of repeats at any other locus), the product rule for probabilities can be applied. This means that if someone has the DNA type of ABC, where the three loci were independent, we can say that the probability of having that DNA type is the probability of having type A times the probability of having type B times the probability of having type C. This has resulted in the ability to generate match probabilities of 1 in a quintillion (1 with 18 zeros after it) or more.

However, DNA database searches showed much more frequent than expected false DNA matches including one perfect 13 locus match out of only 30,000 DNA samples in Maryland in January 2007.

Moreover, since there are about 12 million monozygotic twins on Earth, that theoretical probability is useless. For example, the actual probability that 2 random people have the same DNA depends on whether there were twins or triplets (etc.) in the family, and the number of loci used in the test. Where twins are common, the probability of matching the DNA is 22 in 1000, or about 2.2 in 100 will have matching DNA.

In practice, the risk of contaminated-matching is much greater than matching a distant relative, such as a sample being contaminated from nearby objects, or from left-over cells transferred from a prior test. Logically, the risk is greater for matching the most common person in the samples: everything collected from, or in contact with, a victim is a major source of contamination for any other samples brought into a lab. For that reason, multiple control-samples are typically tested, to ensure that they stayed clean, when prepared during the same period as the actual test samples. Unexpected matches (or variations) in several control-samples indicates a high probability of contamination for the actual test samples. In a relationship test, the full DNA profiles should differ (except for twins), to prove that a person wasn't actually matched as being related to their own DNA in another sample.

AmpFLP

Another technique, AmpFLP, or amplified fragment length polymorphism was also put into practice during the early 1990s. This technique was also faster than RFLP analysis and used PCR to amplify DNA samples. It relied on variable number tandem repeat (VNTR) polymorphisms to distinguish various alleles, which were separated on a polyacrylamide gel using an allelic ladder (as opposed to a molecular weight ladder). Bands could be visualized by silver staining the gel. One popular locus for fingerprinting was the D1S80 locus.

As with all PCR based methods, highly degraded DNA or very small amounts of DNA may cause allelic dropout (causing a mistake in thinking a heterozygote is a homozygote) or other stochastic effects. In addition, because the analysis is done on a gel, very high number repeats may bunch together at the top of the gel, making it difficult to resolve. AmpFLP analysis can be highly automated, and allows for easy creation of phylogenetic trees based on comparing individual samples of DNA. Due to its relatively low cost and ease of set-up and operation, AmpFLP remains popular in lower income countries.

DNA Family Relationship Analysis

Using PCR technology, DNA analysis is widely applied to determine genetic family relationships such as paternity, maternity, siblingship and other kinships.

During conception, the father's sperm cell and the mother's egg cell, each containing half the amount of DNA found in other body cells, meet and fuse to form a fertilized egg, called a zygote. The zygote contains a complete set of DNA molecules, a unique combination of DNA from both parents. This zygote divides and multiplies into an embryo and later, a full human being.

At each stage of development, all the cells forming the body contain the same DNA—half from the father and half from the mother. This fact allows the relationship testing to use all types of all samples including loose cells from the cheeks collected using buccal swabs, blood or other types of samples.

While a lot of DNA contains information for a certain function, there is some called junk DNA, which is currently used for human identification. At some special locations (called loci) in the junk DNA, predictable inheritance patterns were found to be useful in determining biological relationships. These locations contain specific DNA markers that DNA scientists use to identify individuals. In a routine DNA paternity test, the markers used are Short Tandem Repeats (STRs), short pieces of DNA that occur in highly differential repeat patterns among individuals.

Each person's DNA contains two copies of these markers—one copy inherited from the father and one from the mother. Within a population, the markers at each person's DNA location could differ in length and sometimes sequence, depending on the markers inherited from the parents.

The combination of marker sizes found in each person makes up his/her unique genetic profile. When determining the relationship between two individuals, their genetic profiles are compared to see if they share the same inheritance patterns at a statistically conclusive rate.

For example, the following sample report from this commercial DNA paternity testing laboratory Universal Genetics signifies how relatedness between parents and child is identified on those special markers:

DNA Marker	Mother	Child	Alleged father
D21S11	28, 30	28, 31	29, 31
D7S820	9, 10	10, 11	11, 12
TH01	14, 15	14, 16	15, 16
D13S317	7, 8	7, 9	8, 9
D19S433	14, 16.2	14, 15	15, 17

The partial results indicate that the child and the alleged father's DNA match among these five markers. The complete test results show this correlation on 16 markers between the child and the tested man to draw a conclusion of whether or not the man is the biological father.

Scientifically, each marker is assigned with a Paternity Index (PI), which is a statistical measure of how powerfully a match at a particular marker indicates paternity. The PI of each marker is multiplied with each other to generate the Combined Paternity Index (CPI), which indicates the overall probability of an individual being the biological father of the tested child relative to any random man from the entire population of the same race. The CPI is then converted into a Probability of Paternity showing the degree of relatedness between the alleged father and child.

The DNA test report in other family relationship tests, such as grandparentage and siblingship tests, is similar to a paternity test report. Instead of the Combined Paternity Index, a different value, such as a Siblingship Index, is reported.

The report shows the genetic profiles of each tested person. If there are markers shared among the tested individuals, the probability of biological relationship is calculated to determine how likely the tested individuals share the same markers due to a blood relationship.

Y-chromosome Analysis

Recent innovations have included the creation of primers targeting polymorphic regions on the Y-chromosome (Y-STR), which allows resolution of a mixed DNA sample from a male and female and/or cases in which a differential extraction is not possible. Y-chromosomes are paternally inherited, so Y-STR analysis can help in the identification of paternally related males. Y-STR analysis was performed in the Sally Hemings controversy to determine if Thomas Jefferson had sired a son with one of his slaves.

Mitochondrial Analysis

For highly degraded samples, it is sometimes impossible to get a complete profile of the 13 CODIS STRs. In these situations, mitochondrial DNA (mtDNA) is sometimes typed due to there being many copies of mtDNA in a cell, while there may only be 1-2 copies of the nuclear DNA.

Forensic scientists amplify the HV1 and HV2 regions of the mtDNA, then sequence each region and compare single-nucleotide differences to a reference. Because mtDNA is maternally inherited, directly linked maternal relatives can be used as match references, such as one's maternal grandmother's daughter's son.

A difference of two or more nucleotides is generally considered to be an exclusion. Heteroplasmy and poly-C differences may throw off straight sequence comparisons, so some expertise on the part of the analyst is required. mtDNA is useful in determining clear identities, such as those of missing people when a maternally linked relative can be found. mtDNA testing was used in determining that Anna Anderson was not the Russian princess she had claimed to be, Anastasia Romanov.

mtDNA can be obtained from such material as hair shafts and old bones/teeth..

DNA Databases

There are now several DNA databases in existence around the world. Some are private, but most of the largest databases are government controlled. The United States maintains the largest DNA database, with the Combined DNA Index System, holding over 5 million records as of 2007.

The United Kingdom maintains the National DNA Database (NDNAD), which is of similar size, despite the UK's smaller population. The size of this database, and its rate of growth, is giving concern to civil liberties groups in the UK, where police have wide-ranging powers to take samples and retain them even in the event of acquittal.

The U.S. Patriot Act of the United States provides a means for the U.S. government to get DNA samples from other countries if they are either a division of, or head office of, a company operating in the

U.S. Under the act, the American offices of the company can't divulge to their subsidiaries/offices in other countries the reasons that these DNA samples are sought or by whom.

When a match is made from a National DNA Databank to link a crime scene to an offender who has provided a DNA Sample to a databank that link is often referred to as a *cold hit*. A cold hit is of value in referring the police agency to a specific suspect but is of less evidential value than a DNA match made from outside the DNA Databank.

Considerations when Evaluating DNA Evidence

In the early days of the use of genetic fingerprinting as criminal evidence, juries were often swayed by spurious statistical arguments by defence lawyers along these lines: given a match that had a 1 in 5 million probability of occurring by chance, the lawyer would argue that this meant that in a country of say 60 million people there were 12 people who would also match the profile. This was then translated to a 1 in 12 chance of the suspect being the guilty one.

This argument is not sound unless the suspect was drawn at random from the population of the country. In fact, a jury should consider how likely it is that an individual matching the genetic profile would also have been a suspect in the case for other reasons. Another spurious statistical argument is based on the false assumption that a 1 in 5 million probability of a match automatically translates into a 1 in 5 million probability of innocence and is known as the prosecutor's fallacy.

When using RFLP, the theoretical risk of a coincidental match is 1 in 100 billion (100,000,000,000), although the practical risk is actually 1 in 1000 because monozygotic twins are 0.2% of the human population. Moreover, the rate of laboratory error is almost certainly higher than this, and often actual laboratory procedures do not reflect the theory under which the coincidence probabilities were computed.

For example, the coincidence probabilities may be calculated based on the probabilities that markers in two samples have bands in *precisely* the same location, but a laboratory worker may conclude that similar—but not precisely identical—band patterns result from identical genetic samples with some imperfection in the agarose gel. However, in this case, the laboratory worker increases the coincidence risk by expanding the criteria for declaring a match.

Recent studies have quoted relatively high error rates which may be cause for concern. In the early days of genetic fingerprinting, the necessary population data to accurately compute a match probability was sometimes unavailable. Between 1992 and 1996, arbitrary low ceilings were controversially put on match probabilities used in RFLP analysis rather than the higher theoretically computed ones. Today, RFLP has become widely disused due to the advent of more discriminating, sensitive and easier technologies.

STRs do not suffer from such subjectivity and provide similar power of discrimination (1 in 10^{13} for unrelated individuals if using a full SGM+ profile) It should be noted that figures of this magnitude are not considered to be statistically supportable by scientists in the UK, for unrelated individuals with full matching DNA profiles a match probability of 1 in a billion is considered statistically supportable (Since 1998 the DNA profiling system supported by The National DNA Database in the UK is the SGM+ DNA profiling system which includes 10 STR regions and a sex indicating test. However, with any DNA technique, the cautious juror should not convict on genetic fingerprint evidence alone if other factors raise doubt. Contamination with other evidence (secondary transfer) is a key source of incorrect DNA profiles and raising doubts as to whether a sample has been adulterated is a favorite defence technique. More rarely, chimerism is one such instance where the lack of a genetic match may unfairly exclude a suspect.

Evidence of Genetic Relationship

It's also possible to use DNA profiling as evidence of genetic relationship, but testing that shows no relationship isn't absolutely certain. While almost all individuals have a single and distinct set of genes, rare individuals, known as "chimeras", have at least two different sets of genes. There have been several cases of DNA profiling that falsely "proved" that a mother was unrelated to her children.

Fake DNA Evidence

The value of DNA evidence has to be seen in light of recent cases where criminals planted fake DNA samples at crime scenes. In one case, a criminal even planted fake DNA evidence in his own body: Dr. John Schneeberger raped one of his sedated patients in 1992 and left semen on her underwear. Police drew what they believed to be Schneeberger's blood and compared its DNA against the crime scene semen DNA on three occasions, never showing a match. It turned out that he had surgically inserted a Penrose drain into his arm and filled it with foreign blood and anticoagulants.

In a study conducted by the life science company Nucleix and published in the journal Forensic Science International, scientists found that an In vitro synthesized sample of DNA matching any desired genetic profile can be constructed using standard molecular biology techniques without obtaining any actual tissue from that person.

DNA Evidence as Evidence in Criminal Trials

Familial Searching

Familial searching is the use of family members' DNA to identify a closely related suspect in jurisdictions where large DNA databases exist, but no exact match has been found. The first successful use of the practice was in a UK case where a man was convicted of manslaughter when he threw a brick stained with his own blood into a moving car.

Police could not get an exact match to the UK's DNA database because the man had no criminal convictions, but police implicated him using a close relative's DNA. The technique was used to catch a Los Angeles serial killer known as the "Grim Sleeper" in 2010. However, critics have challenged the technology as "deeply antithetical to American values" and likely to lead to excess investigation of racial and ethnic minorities.

Surreptitious DNA Collecting

Police forces may collect DNA samples without the suspects' knowledge, and use it as evidence. Legality of this mode of proceeding has been questioned in Australia. In the United States, it has been accepted, courts often claiming that there was no expectation of privacy, citing California v. Greenwood (1985), during which the Supreme Court held that the Fourth Amendment does not prohibit the warrantless search and seizure of garbage left for collection outside the curtilage of a home. Critics of this practice underline the fact that this analogy ignores that "most people have no idea that they risk surrendering their genetic identity to the police by, for instance, failing to destroy a used coffee cup.

Moreover, even if they do realize it, there is no way to avoid abandoning one's DNA in public." In the UK, the Human Tissue Act

2004 prohibited private individuals from covertly collecting biological samples (hair, fingernails, etc.) for DNA analysis, but excluded medical and criminal investigations from the offense.

England and Wales

Evidence from an expert who has compared DNA samples must be accompanied by evidence as to the sources of the samples and the procedures for obtaining the DNA profiles. The judge must ensure that the jury must understand the significance of DNA matches and mismatches in the profiles. The judge must also ensure that the jury does not confuse the 'match probability' (the probability that a person that is chosen at random has a matching DNA profile to the sample from the scene) with the 'likelihood ratio' (the probability that a person with matching DNA committed the crime). In *R v. Doheny*, EWCA Crim 728. Phillips LJ gave this example of a summing up, which should be carefully tailored to the particular facts in each case:

Members of the Jury, if you accept the scientific evidence called by the Crown, this indicates that there are probably only four or five white males in the United Kingdom from whom that semen stain could have come. The Defendant is one of them. If that is the position, the decision you have to reach, on all the evidence, is whether you are sure that it was the Defendant who left that stain or whether it is possible that it was one of that other small group of men who share the same DNA characteristics.

Juries should weigh up conflicting and corroborative evidence, using their own common sense and not by using mathematical formulae, such as Bayes' theorem, so as to avoid "confusion, misunderstanding and misjudgment".

Presentation and Evaluation of Evidence of Partial or Incomplete DNA Profiles

R v Bates (2006) EWCA Crim 1395 Moore-Bick LJ said: "We can see no reason why partial profile DNA evidence should not be admissible provided that the jury are made aware of its inherent limitations and are given a sufficient explanation to enable them to evaluate it.

There may be cases where the match probability in relation to all the samples tested is so great that the judge would consider its probative value to be minimal and decide to exclude the evidence in the exercise of his discretion, but this gives rise to no new question of principle and can be left for decision on a case by case basis.

However, the fact that there exists in the case of all partial profile evidence the possibility that a "missing" allele might exculpate the accused altogether does not provide sufficient grounds for rejecting such evidence. In many there is a possibility (at least in theory) that evidence exists which would assist the accused and perhaps even exculpate him altogether, but that does not provide grounds for excluding relevant evidence that is available and otherwise admissible, though it does make it important to ensure that the jury are given sufficient information to enable them to evaluate that evidence properly".

DNA Testing in the US

There are state laws on DNA profiling in all 50 states of the United States. Detailed information on database laws in each state can be found at the National Conference of State Legislatures website.

Development of Artificial DNA

In August 2009, scientists in Israel stunned the forensic sciences and raised serious questions concerning the use of DNA by law enforcement as the ultimate method of identification. In a paper published in the journal *Forensic Science International: Genetics*, the Israeli researchers demonstrated that it is possible to manufacture DNA in a laboratory, and thus falsify DNA evidence. The scientists fabricated saliva and blood samples, which originally contained DNA from a person other than the ostensible donor of the blood and saliva.

The researchers also showed that, using a DNA database, it is possible to take information from a profile and manufacture DNA to match it, and that this can be done without access to any actual DNA from the person whose DNA they are duplicating. The synthetic DNA oligos required for the procedure are common in molecular laboratories.

Dr. Daniel Frumkin, lead author on the paper, was quoted in *The New York Times* as saying, "You can just engineer a crime scene... any biology undergraduate could perform this."

Dr. Frumkin perfected a test that can forensically differentiate real DNA samples from fake ones. His test detects epigenetic modifications, in particular, DNA methylation. Seventy percent of the DNA in any human genome is methylated, meaning it contains methyl group modifications within a CpG dinucleotide context. Methylation at the promoter region is associated with gene silencing. The synthetic DNA lacks this epigenetic modification, which allows the test to distinguish manufactured DNA from original, genuine, DNA.

It is unknown how many, if any, police departments currently use the test, which appears to be a serious issue in light of Frumkin's claim that the DNA manufacturing procedure is within the grasp of any undergraduate biology student. No police lab has publicly announced that it is using the new test to verify DNA results, while FSI Genetics says that any forensic laboratory doing DNA identification should adopt this test to authenticate its results as "real" DNA.

Bioinformatics

Bioinformatics involves the manipulation, searching, and data mining of DNA sequence data. The development of techniques to store and search DNA sequences have led to widely applied advances in computer science, especially string searching algorithms, machine learning and database theory. String searching or matching algorithms, which find an occurrence of a sequence of letters inside a larger sequence of letters, were developed to search for specific sequences of nucleotides. In other applications such as text editors, even simple algorithms for this problem usually suffice, but DNA sequences cause these algorithms to exhibit near-worst-case behaviour due to their small number of distinct characters.

The related problem of sequence alignment aims to identify homologous sequences and locate the specific mutations that make them distinct. These techniques, especially multiple sequence alignment, are used in studying phylogenetic relationships and protein function.

Data sets representing entire genomes' worth of DNA sequences, such as those produced by the Human Genome Project, are difficult to use without annotations, which label the locations of genes and regulatory elements on each chromosome. Regions of DNA sequence that have the characteristic patterns associated with protein- or RNA-coding genes can be identified by gene finding algorithms, which allow researchers to predict the presence of particular gene products in an organism even before they have been isolated experimentally.

DNA Nanotechnology

DNA nanotechnology uses the unique molecular recognition properties of DNA and other nucleic acids to create self-assembling branched DNA complexes with useful properties. DNA is thus used as a structural material rather than as a carrier of biological information.

This has led to the creation of two-dimensional periodic lattices (both tile-based as well as using the "DNA origami" method) as well as three-dimensional structures in the shapes of polyhedra. Nanomechanical devices and algorithmic self-assembly have also been demonstrated, and these DNA structures have been used to template the arrangement of other molecules such as gold nanoparticles and streptavidin proteins.

History and Anthropology

Because DNA collects mutations over time, which are then inherited, it contains historical information, and, by comparing DNA sequences, geneticists can infer the evolutionary history of organisms, their phylogeny. This field of phylogenetics is a powerful tool in evolutionary biology. If DNA sequences within a species are compared, population geneticists can learn the history of particular populations. This can be used in studies ranging from ecological genetics to anthropology; For example, DNA evidence is being used to try to identify the Ten Lost Tribes of Israel.

DNA has also been used to look at modern family relationships, such as establishing family relationships between the descendants of Sally Hemings and Thomas Jefferson. This usage is closely related to the use of DNA in criminal investigations detailed above. Indeed, some criminal investigations have been solved when DNA from crime scenes has matched relatives of the guilty individual.

History of DNA Research

DNA was first isolated by the Swiss physician Friedrich Miescher who, in 1869, discovered a microscopic substance in the pus of discarded surgical bandages. As it resided in the nuclei of cells, he called it "nuclein". In 1919, Phoebus Levene identified the base, sugar and phosphate nucleotide unit. Levene suggested that DNA consisted of a string of nucleotide units linked together through the phosphate groups. However, Levene thought the chain was short and the bases repeated in a fixed order. In 1937 William Astbury produced the first X-ray diffraction patterns that showed that DNA had a regular structure.

In 1928, Frederick Griffith discovered that traits of the "smooth" form of the *Pneumococcus* could be transferred to the "rough" form of the same bacteria by mixing killed "smooth" bacteria with the live "rough" form. This system provided the first clear suggestion that

DNA carries genetic information—the Avery-MacLeod-McCarty experiment—when Oswald Avery, along with coworkers Colin MacLeod and Maclyn McCarty, identified DNA as the transforming principle in 1943. DNA's role in heredity was confirmed in 1952, when Alfred Hershey and Martha Chase in the Hershey-Chase experiment showed that DNA is the genetic material of the T2 phage.

In 1953, James D. Watson and Francis Crick suggested what is now accepted as the first correct double-helix model of DNA structure in the journal *Nature*. Their double-helix, molecular model of DNA was then based on a single X-ray diffraction image (labeled as "Photo 51") taken by Rosalind Franklin and Raymond Gosling in May 1952, as well as the information that the DNA bases are paired — also obtained through private communications from Erwin Chargaff in the previous years. Chargaff's rules played a very important role in establishing double-helix configurations for B-DNA as well as A-DNA.

Experimental evidence supporting the Watson and Crick model were published in a series of five articles in the same issue of *Nature*. Of these, Franklin and Gosling's paper was the first publication of their own X-ray diffraction data and original analysis method that partially supported the Watson and Crick model; this issue also contained an article on DNA structure by Maurice Wilkins and two of his colleagues, whose analysis and *in vivo* B-DNA X-ray patterns also supported the presence *in vivo* of the double-helical DNA configurations as proposed by Crick and Watson for their double-helix molecular model of DNA in the previous two pages of *Nature*. In 1962, after Franklin's death, Watson, Crick, and Wilkins jointly received the Nobel Prize in Physiology or Medicine. However, Nobel rules of the time allowed only living recipients, but a vigorous debate continues on who should receive credit for the discovery.

In an influential presentation in 1957, Crick laid out the "Central Dogma" of molecular biology, which foretold the relationship between DNA, RNA, and proteins, and articulated the "adaptor hypothesis". Final confirmation of the replication mechanism that was implied by the double-helical structure followed in 1958 through the Meselson-Stahl experiment. Further work by Crick and coworkers showed that the genetic code was based on non-overlapping triplets of bases, called codons, allowing Har Gobind Khorana, Robert W. Holley and Marshall Warren Nirenberg to decipher the genetic code. These findings represent the birth of molecular biology.

Protein Synthesis

Protein synthesis is the process in which cells build proteins. The term is sometimes used to refer only to protein translation but more often it refers to a multi-step process, beginning with amino acid synthesis and transcription of nuclear DNA into messenger RNA, which is then used as input to translation.

The cistron DNA is transcribed into a variety of RNA intermediates. The last version is used as a template in synthesis of a polypeptide chain. Proteins can often be synthesized directly from genes by translating mRNA. When a protein needs to be available on short notice or in large quantities, a protein precursor is produced. A proprotein is an inactive protein containing one or more inhibitory peptides that can be activated when the inhibitory sequence is removed by proteolysis during posttranslational modification.

A preprotein is a form that contains a signal sequence (an N-terminal signal peptide) that specifies its insertion into or through membranes; i.e., targets them for secretion. The signal peptide is cleaved off in the endoplasmic reticulum. Preproproteins have both sequences (inhibitory and signal) still present.

For synthesis of protein, a succession of tRNA molecules charged with appropriate amino acids have to be brought together with an mRNA molecule and matched up by base-pairing through their anti-codons with each of its successive codons. The amino acids then have to be linked together to extend the growing protein chain, and the tRNAs, relieved of their burdens, have to be released. This whole complex of processes is carried out by a giant multimolecular machine, the ribosome, formed of two main chains of RNA, called ribosomal RNA (rRNA), and more than 50 different proteins.

This molecular juggernaut latches onto the end of an mRNA molecule and then trundles along it, capturing loaded tRNA molecules and stitching together the amino acids they carry to form a new protein chain. Protein biosynthesis, although very similar, is different for prokaryotes and eukaryotes.

Amino Acid Synthesis

Amino acid synthesis is the set of biochemical processes (metabolic pathways) by which the various amino acids are produced from other compounds. The substrates for these processes are various compounds in the organism's diet or growth media. Not all organisms are able to synthesise all amino acids, for example humans are only able to synthesise 12 of the 20 standard amino acids.

A fundamental problem for biological systems is to obtain nitrogen in an easily usable form. This problem is solved by certain microorganisms capable of reducing the inert Na"N molecule (nitrogen gas) to two molecules of ammonia in one of the most remarkable reactions in biochemistry.

Nitrogen in the form of ammonia is the source of nitrogen for all the amino acids. The carbon backbones come from the glycolytic pathway, the pentose phosphate pathway, or the citric acid cycle.

In amino acid production, one encounters an important problem in biosynthesis— namely, stereochemical control. Because all amino acids except glycine are chiral, biosynthetic pathways must generate the correct isomer with high fidelity.

In each of the 19 pathways for the generation of chiral amino acids, the stereochemistry at the á-carbon atom is established by a transamination reaction that involves pyridoxal phosphate. Almost all the transaminases that catalyze these reactions descend from a common ancestor, illustrating once again that effective solutions to biochemical problems are retained throughout evolution.

Biosynthetic pathways are often highly regulated such that building blocks are synthesized only when supplies are low. Very often, a high concentration of the final product of a pathway inhibits the activity of enzymes that function early in the pathway. Often present are allosteric enzymes capable of sensing and responding to concentrations of regulatory species. These enzymes are similar in functional properties to aspartate transcarbamoylase and its regulators. Feedback and allosteric mechanisms ensure that all twenty amino acids are maintained in sufficient amounts for protein synthesis and other processes.

Amino acids are synthesized from α-ketoacids, and later transaminated from another aminoacid, usually Glutamate. The enzyme involved in this reaction is an aminotransferase.

Glutamate itself is formed by amination of α-ketoglutarate:

Nitrogen Fixation: Microorganisms Use ATP and a Powerful Reductant to Reduce Atmospheric Nitrogen to Ammonia

Microorganisms use ATP and reduced ferredoxin, a powerful reductant, to reduce N2 to NH3. An iron-molybdenum cluster in nitrogenase deftly catalyzes the fixation of N2, a very inert molecule. Higher organisms consume the fixed nitrogen to synthesize amino acids, nucleotides, and other nitrogen-containing biomolecules. The major points of entry of NH4+ into metabolism are glutamine or glutamate.

Amino Acids are Made from Intermediates of the Citric Acid Cycle and Other Major Pathways

Of the basic set of 20 amino acids (not counting selenocysteine) there are 8 that human beings cannot synthesize. In addition, the amino acids arginine, cysteine, glycine, glutamine, histidine, proline, serine and tyrosine are considered conditionally essential, meaning they are not normally required in the diet, but must be supplied exogenously to specific populations that do not synthesize it in adequate amounts. (For example, enough arginine is synthesized by the urea cycle to meet the needs of an adult but perhaps not those of a growing child.)

Amino acids that need to be obtained from the diet are called essential amino acids. Nonessential amino acids are produced in the body. The pathways for the synthesis of nonessential amino acids are quite simple. Glutamate dehydrogenase catalyzes the reductive amination of α-ketoglutarate to glutamate. A transamination reaction takes place in the synthesis of most amino acids. At this step, the chirality of the amino acid is established. Alanine and aspartate are synthesized by the transamination of pyruvate and oxaloacetate, respectively. Glutamine is synthesized from NH4+ and glutamate, and asparagine is synthesized similarly. Proline and arginine are derived from glutamate.

Serine, formed from 3-phosphoglycerate, is the precursor of glycine and cysteine. Tyrosine is synthesized by the hydroxylation of phenylalanine, an essential amino acid. The pathways for the biosynthesis of essential amino acids are much more complex than those for the nonessential ones.

Tetrahydrofolate, a carrier of activated one-carbon units, plays an important role in the metabolism of amino acids and nucleotides. This coenzyme carries one-carbon units at three oxidation states, which are interconvertible: most reduced—methyl; intermediate—methylene; and most oxidized—formyl, formimino, and methenyl. The major donor of activated methyl groups is S-adenosylmethionine, which is synthesized by the transfer of an adenosyl group from ATP to the sulfur atom of methionine. S-Adenosylhomocysteine is formed when the activated methyl group is transferred to an acceptor. It is hydrolyzed to adenosine and homocysteine, the latter of which is then methylated to methionine to complete the activated methyl cycle. The A Cortisol inhibites protein synthesis.

Amino Acid Biosynthesis is Regulated by Feedback Inhibition

Most of the pathways of amino acid biosynthesis are regulated by feedback inhibition, in which the committed step is allosterically inhibited by the final product. Branched pathways require extensive interaction among the branches that includes both negative and positive regulation. The regulation of glutamine synthetase from E. coli is a striking demonstration of cumulative feedback inhibition and of control by a cascade of reversible covalent modifications.

Amino Acids are Precursors of Many Biomolecules

Amino acids are precursors of a variety of biomolecules. Glutathione serves as a sulfhydryl buffer and detoxifying agent. Glutathione peroxidase, a selenoenzyme, catalyzes the reduction of hydrogen peroxide and organic peroxides by glutathione. Nitric oxide, a short-lived messenger, is formed from arginine. Porphyrins are synthesized from glycine and succinyl CoA, which condense to give β-aminolevulinate. Two molecules of this intermediate become linked to form porphobilinogen. Four molecules of porphobilinogen combine to form a linear tetrapyrrole, which cyclizes to uroporphyrinogen III. Oxidation and side-chain modifications lead to the synthesis of protoporphyrin IX, which acquires an iron atom to form heme.

Transcription

In transcription an mRNA chain is generated, with one strand of the DNA double helix in the genome as template. This strand is called the template strand. Transcription can be divided into 3 stages: Initiation, Elongation and Termination, each regulated by a large number of proteins such as transcription factors and coactivators that ensure the correct gene is transcribed.

The DNA strand is read in the 3' to 5' direction and the mRNA is transcribed in the 5' to 3' direction by the RNA polymerase.

Transcription occurs in the cell nucleus, where the DNA is held. The DNA structure of the cell is made up of two helixes made up of sugar and phosphate held together by the bases. The sugar and the phosphate are joined together by covalent bond. The DNA is "unzipped" by the enzyme helicase, leaving the single nucleotide chain open to be copied. RNA polymerase reads the DNA strand from 3 prime (3') end to the 5 prime (5') end, while it synthesizes a single strand of messenger RNA in the 5' to 3' direction. The general RNA structure is very similar to the DNA structure, but in RNA the nucleotide uracil takes the place that thymine occupies in DNA. The single strand of mRNA leaves the nucleus through nuclear pores, and migrates into the cytoplasm.

The first product of transcription differs in prokaryotic cells from that of eukaryotic cells, as in prokaryotic cells the product is mRNA, which needs no post-transcriptional modification, while in eukaryotic cells, the first product is called primary transcript, that needs post-transcriptional modification (capping with 7 methyl guanosine, tailing with a poly A tail) to give hnRNA (heterophil nuclear RNA). hnRNA then undergoes splicing of introns (non coding parts of the gene) via spliceosomes to produce the final mRNA.

Translation

The synthesis of proteins is known as translation. Translation occurs in the cytoplasm, where the ribosomes are located. Ribosomes are made of a small and large subunit that surround the mRNA. In translation, messenger RNA (mRNA) is decoded to produce a specific polypeptide according to the rules specified by the trinucleotide genetic code.

This uses an mRNA sequence as a template to guide the synthesis of a chain of amino acids that form a protein. Translation proceeds in four phases: activation, initiation, elongation, and termination (all describing the growth of the amino acid chain, or polypeptide that is the product of translation).

In activation, the correct amino acid (AA) is joined to the correct transfer RNA (tRNA). While this is not technically a step in translation, it is required for translation to proceed. The AA is joined by its carboxyl group to the 3' OH of the tRNA by an ester bond.

When the tRNA has an amino acid linked to it, it is termed "charged". Initiation involves the small subunit of the ribosome binding to 5' end of mRNA with the help of initiation factors (IF), other proteins that assist the process. Elongation occurs when the next aminoacyl-tRNA (charged tRNA) in line binds to the ribosome along with GTP and an elongation factor.

Termination of the polypeptide happens when the A site of the ribosome faces a stop codon (UAA, UAG, or UGA). When this happens, no tRNA can recognize it, but releasing factor can recognize nonsense codons and causes the release of the polypeptide chain. The capacity of disabling or inhibiting translation in protein biosynthesis is used by antibiotics such as: anisomycin, cycloheximide, chloramphenicol, tetracycline, streptomycin, erythromycin, puromycin etc.

Translation is the process of converting the mRNA codon sequences into an amino acid polypeptide chain.

1. Amino acid activation
2. Initiation - A ribosome attaches to the mRNA and starts to code at the FMet codon (usually AUG, sometimes GUG or UUG).
3. Elongation - tRNA brings the corresponding amino acid (which has an anticodon that identifies the amino acid as the corresponding molecule to a codon) to each codon as the ribosome moves down the mRNA strand.
4. Termination - Reading of the final mRNA codon (aka the STOP codon), which ends the synthesis of the peptide chain and releases it.

Events Following Protein Translation

The events following biosynthesis include post-translational modification and protein folding. During and after synthesis, polypeptide chains often fold to assume, so called, native secondary and tertiary structures. This is known as *protein folding*.

Many proteins undergo *post-translational modification*. This may include the formation of disulfide bridges or attachment of any of a number of biochemical functional groups, such as acetate, phosphate, various lipids and carbohydrates. Enzymes may also remove one or more amino acids from the leading (amino) end of the polypeptide chain, leaving a protein consisting of two polypeptide chains connected by disulfide bonds.

In general, protein molecules are believed to be modified by small chemical groups, post-translationally. Chemical modifications such as, phosphorylation of serine/threonine, acetylation or methylation of lysine, hydroxylation of proline/lysine, formylation of glycine, glycosylation of serine/threonine/asparagine, acylation of cysteine, myristoylation of glycine, biotinylation of lysine, ubiquitination, etc. on proteins is a very important issue in relation to properly understanding the biological functions of a given protein. These much-studied post-translational modifications have become well-established with the discovery of the respective enzymes (kinases for phosphorylation, acetylases for acetylation, methyl transferases for methylation, etc.), which carry on the chemical modifications on the specific amino acid residues. All of these modifications are still believed to have happened as post-translational events. There is no study yet on when actually one particular modification occurs on a given amino acid residue in a given protein. Does it happen when the protein is already formed, or when the amino acid chain is being synthesized, or before the translation of the primary chain has begun?

Since these chemical modifications are related to the biological functions of a protein, it is easy to think that these chemical modifications have happened to the whole protein molecule, after the protein primary chain is fully synthesized; but, if that is the case, we have to consider the fact that the primary chains get folded instantly, (in a similar way as the newly synthesized DNA strands form helixes), to attain its compact-globular conformation ; As most of the primary chains are fairly long [a 5Kd protein may have 40-45 amino acid residues in its primary chain], it is likely that the newly formed amino acid chain tries to remain intact by folding, thereby avoiding its breakdown via lots of proteases present within the cytoplasm. And, no capping event to protect the N-terminal end of the primary sequence (similar to 5' m-RNA capping to protect m-RNAs) is ever discovered for protein primary structure. So, by folding mechanism, the primary chain, perhaps, avoids the protease attacks.

However, once it gets folded, it may be very difficult for the respective enzyme molecule to find out the particular aa residue from the complexity of that compactly folded conformation. In addition, it can be clearly imagined that this enzymatic modification/reaction on a given amino acid requires presence and association of the appropriate enzyme, necessary cofactors, etc. This association is much easier to occur when the amino acid residues in the primary structure are readily available for binding; in other words, it is much difficult for the enzyme molecules to find and to bind to its substrate amino acid residue in a mature protein molecule after its three-dimensional conformation been attained.

So one can think that the modifications can happen while the primary chain of the protein is being synthesized during the translation process on the m-RNA strand; the amino acid residues on the primary chain can be modified instantly and enzymatically by kinases, acetylases, hydroxylases, methyltransferases, etc. to initiate proper folding for protection in order to avoid degradation by proteases, thereby gaining the globular form.

Also, the reader can imagine another scenario in which, while the free amino acid molecules are formed within a cell and become available, they [in that free state] may be modified enzymatically before taking the ride to the translational event; this means that, while the primary chain is being synthesized, the pre-modified amino acid molecules are ready to be engaged. This way, as the primary chain is getting synthesized, it does not have to be modified by the modifying enzymes anymore, and can fold itself instantaneously without concern of degradation. This also arises new thoughts that (i) phosphate, acetyl, methyl, biotinyl, acyl, etc. groups may have the ability to inhibit protease actions on the primary chains; (ii) most or all of the amino acid residues get modified by small chemical groups (so far, only some chemical groups are known).

It is still not fully known exactly when and how the actual modification of a given amino acid residue occurs, at which stage of synthesis, within a protein molecule.

Once the chemically-modified, protease-insensitive, intact protein molecule is generated, it must perform its biological function, which requires its being activated. The activation is probably done by a second set of enzymatic reactions when these chemical groups are removed from the a residues (or added back). So, the second type of post-translational modifications are the opposite reactions of the above-described type, which are dephosphorylation by phosphatases or phosphoryl transferases, deacetylation by deacetylases or acetyltransferases, demethylation by demethylases or methyl-transferases, ubiquitination, SuMoylation, glycosylation, biotinylation, etc. These are taking place on the whole protein molecule toward generating its activated form to execute a particular function or toward its deactivation followed by its total degradation. As the chemical groups are removed (or added), it is quite clear that the proteins can go through different states of structural/conformational change. Thus, after translation, a protein can change its conformation dynamically while the chemical groups are removed or added enzymatically; these conformational changes in the protein structure help the protein to proceed through its lifecycle, until it is ubiquitinated for its total degradation.

┌─────────┐
│   7   │
└─────────┘

Chromosomes

A chromosome is an organized building of DNA and protein that is found in cells. It is a single piece of coiled DNA containing many genes, regulatory elements and other nucleotide sequences. Chromosomes also contain DNA-bound proteins, which serve to package the DNA and control its functions.

Chromosomes vary widely between different organisms. The DNA molecule may be circular or linear, and can be composed of 10,000 to 1,000,000,000 nucleotides in a long chain. Typically eukaryotic cells (cells with nuclei) have large linear chromosomes and prokaryotic cells (cells without defined nuclei) have smaller circular chromosomes, although there are many exceptions to this rule. Furthermore, cells may contain more than one type of chromosome; for example, mitochondria in most eukaryotes and chloroplasts in plants have their own small chromosomes.

In eukaryotes, nuclear chromosomes are packaged by proteins into a condensed structure called chromatin. This allows the very long DNA molecules to fit into the cell nucleus. The structure of chromosomes and chromatin varies through the cell cycle. Chromosomes are the essential unit for cellular division and must be replicated, divided, and passed successfully to their daughter cells so as to ensure the genetic diversity and survival of their progeny. Chromosomes may exist as either duplicated or unduplicated— unduplicated chromosomes are single linear strands, whereas duplicated chromosomes (copied during synthesis phase) contain two copies joined by a centromere. Compaction of the duplicated chromosomes during mitosis and meiosis results in the classic four-arm structure (pictured to the right). Chromosomal recombination plays a vital role in genetic diversity. If these structures are manipulated incorrectly, through processes known as chromosomal instability and translocation, the cell may undergo mitotic catastrophe and die, or it may aberrantly evade apoptosis leading to the progression of cancer.

In practice "chromosome" is a rather loosely defined term. In prokaryotes and viruses, the term genophore is more appropriate when no chromatin is present. However, a large body of work uses the term chromosome regardless of chromatin content. In prokaryotes DNA is usually arranged as a circle, which is tightly coiled in on itself, sometimes accompanied by one or more smaller, circular DNA molecules called plasmids. These small circular genomes are also found in mitochondria and chloroplasts, reflecting their bacterial origins. The simplest genophores are found in viruses: these DNA or RNA molecules are short linear or circular genophores that often lack structural proteins.

History

Chromosomes as Vectors of Heredity

In a series of experiments, Theodor Boveri gave the definitive demonstration that chromosomes are the vectors of heredity. His two principles were based upon the *continuity* of chromosomes and the *individuality* of chromosomes.

It is the second of these principles that was so original. Boveri was able to test the proposal put forward by Wilhelm Roux, that each chromosome carries a different genetic load, and showed that Roux was right. Upon the rediscovery of Mendel, Boveri was able to point out the connection between the rules of inheritance and the behaviour of the chromosomes. It is interesting to see that Boveri influenced two generations of American cytologists: Edmund Beecher Wilson, Walter Sutton and Theophilus Painter were all influenced by Boveri (Wilson and Painter actually worked with him).

In his famous textbook *The Cell*, Wilson linked Boveri and Sutton together by the Boveri-Sutton theory. Mayr remarks that the theory was hotly contested by some famous geneticists: William Bateson, Wilhelm Johannsen, Richard Goldschmidt and T.H. Morgan, all of a rather dogmatic turn-of-mind. Eventually complete proof came from chromosome maps in Morgan's own lab.

Chromosomes in Eukaryotes

Eukaryotes (cells with nuclei such as those found in plants, yeast, and animals) possess multiple large linear chromosomes contained in the cell's nucleus. Each chromosome has one centromere, with one or two arms projecting from the centromere, although, under most circumstances, these arms are not visible as such. In addition, most eukaryotes have a small circular mitochondrial genome, and some eukaryotes may have additional small circular or linear cytoplasmic chromosomes.

In the nuclear chromosomes of eukaryotes, the uncondensed DNA exists in a semi-ordered structure, where it is wrapped around histones (structural proteins), forming a composite material called chromatin.

Eukaryotic Chromosome Fine Structure

Eukaryotic chromosome fine structure refers to the structure of sequences for eukaryotic chromosomes. Some fine sequences are included in more than one class, so the classification listed is not intended to be completely separate.

Chromosomal Characteristics

Some sequences are required for a properly functioning chromosome:

- Centromere: Used during cell division as the attachment point for the spindle fibres.
- Telomere: Used to maintain chromosomal integrity by capping off the ends of the linear chromosomes. This region is a microsatellite, but its function is more specific than a simple tandem repeat.

Structural Sequences

Other sequences are used in replication or during interphase with the physical structure of the chromosome.

- Ori, or Origin: Origins of replication.
- MAR: Matrix attachment regions, where the DNA attaches to the nuclear matrix.

Protein-coding Genes

Regions of the genome with protein-coding genes include several elements:

- Enhancer regions (normally up to a few thousand basepairs upstream of transcription).

- Promoter regions (normally less than a couple of hundred basepairs upstream of transcription) include elements such as the TATA and CAAT boxes, GC elements, an initiator, *etc.*

- Exons are the part of the transcript that will eventually be transported to the cytoplasm for translation. When discussing gene with alternate splicing, an exon is a portion of the transcript that could be translated, given the correct splicing conditions. The exons can be divided into three parts

  — The coding region is the portion of the mRNA that will eventually be translated.

  — Upstream untranslated region (5' UTR) can serve several functions, including mRNA transport, and initiation of translation (including, portions of the Kozak sequence). They are never translated into the protein (excepting various mutations).

  — The 3' region downstream from the stop codon is separated into two parts:

    – 3' UTR is never translated, but serves to add mRNA stability. It is also the attachment site for the poly-A tail. The poly-A tail is used in the initiation of translation and also seems to have an effect on the long-term stability (aging) of the mRNA.

    – An unnamed region after the poly-A tail, but before the actual site for transcription termination, is spliced off during transcription, and so does not become part of the 3' UTR. Its function, if any, is unknown.

- Introns are intervening sequences between the exons that are never translated. Some sequences inside introns function as miRNA, and there are even some cases of small genes residing completely within the intron of a large gene. For some genes (such as the antibody genes), internal control regions are found inside introns. These situations, however, are treated as exceptions.

Genes that are Used as RNA

Many regions of the DNA are transcribed with RNA as the functional form:

- rRNA: Ribosomal RNA are used in the ribosome.
- tRNA: Transfer RNA are used in the translation process by bringing amino acids to the ribosome.
- snRNA: Small nuclear RNA are used in spliceosomes to help the processing of pre-mRNA.
- gRNA: Guide RNA are used in RNA editing.
- miRNA: Micro RNA are small (approximately 24 nucleotides) that are used in gene silencing.
- snoRNA: Small nucleolar RNA are used to help process and construct the ribosome.

Other RNAs are transcribed and not translated, but have undiscovered functions.

Repeated Sequences

Repeated sequences are of two basic types: unique sequences that are repeated in one area; and repeated sequences that are interspersed throughout the genome.

Satellites

Satellites are unique sequences that are repeated in tandem in one area. Depending on the length of the repeat, they are classified as either:

- Minisatellite: Short repeats of nucleotides.
- Microsatellite: Very short repeats of nucleotides. Some trinucleotide repeats are found in coding regions. Most are found in non-coding regions. Their function is unknown, if they have any specific function. They are used as molecular markers and in DNA fingerprinting.

Interspersed Sequences

Interspersed sequences are tandem repeats, with sequences that are found interspersed across the genome. They can be classified based on the length of the repeat as:

- SINE: Short interspersed sequences. The repeats are normally a few hundred base pairs in length. These sequences constitute about 13% of the human genome with the specific *Alu* sequence accounting for 5%.
- LINE: Long interspersed sequences. The repeats are normally several thousand base pairs in length. These sequences constitute about 21% of the human genome.

Retrotransposons

Retrotransposons are sequences in the DNA that are the result of retrotransposition of RNA. LINEs and SINEs are examples where the sequences are repeats, but there are non-repeated sequences that can also be retrotransposons.

Other Sequences

Typical eukaryotic chromosomes contain much more DNA than is classified in the categories above. The DNA may be used as spacing, or have other as-yet-unknown function. Or, they may simply be random sequences of no consequence.

Chromatin

Chromatin is the complex of DNA and protein found in the eukaryotic nucleus, which packages chromosomes. The structure of chromatin varies significantly between different stages of the cell cycle, according to the requirements of the DNA.

Interphase Chromatin

During interphase (the period of the cell cycle where the cell is not dividing), two types of chromatin can be distinguished:

- Euchromatin, which consists of DNA that is active, e.g., being expressed as protein.
- Heterochromatin, which consists of mostly inactive DNA. It seems to serve structural purposes during the chromosomal stages. Heterochromatin can be further distinguished into two types:
  — *Constitutive heterochromatin*, which is never expressed. It is located around the centromere and usually contains repetitive sequences.
  — *Facultative heterochromatin*, which is sometimes expressed.

Individual chromosomes cannot be distinguished at this stage – they appear in the nucleus as a homogeneous tangled mix of DNA and protein.

Metaphase Chromatin and Division

In the early stages of mitosis or meiosis (cell division), the chromatin strands become more and more condensed. They cease to function as accessible genetic material (transcription stops) and become a compact transportable form. This compact form makes the individual chromosomes visible, and they form the classic four arm structure, a pair of sister chromatids attached to each other at the centromere. The shorter arms are called *p arms* (from the French *petit*, small) and the longer arms are called *q arms* (*q* follows *p* in the Latin alphabet). This is the only natural context in which individual chromosomes are visible with an optical microscope.

During divisions, long microtubules attach to the centromere and the two opposite ends of the cell. The microtubules then pull the chromatids apart, so that each daughter cell inherits one set of chromatids. Once the cells have divided, the chromatids are uncoiled and can function again as chromatin. In spite of their appearance, chromosomes are structurally highly condensed, which enables these giant DNA structures to be contained within a cell nucleus.

The self-assembled microtubules form the spindle, which attaches to chromosomes at specialized structures called kinetochores, one of which is present on each sister chromatid. A special DNA base sequence in the region of the kinetochores provides, along with special proteins, longer-lasting attachment in this region.

Mitosis

Mitosis is the process by which a eukaryotic cell separates the chromosomes in its cell nucleus into two identical sets in two nuclei. It is generally followed immediately by cytokinesis, which divides the nuclei, cytoplasm, organelles and cell membrane into two cells containing roughly equal shares of these cellular components. Mitosis and cytokinesis together define the mitotic (M) phase of the cell cycle - the division of the mother cell into two daughter cells, genetically identical to each other and to their parent cell. This accounts for approximately 10% of the cell cycle.

Mitosis occurs exclusively in eukaryotic cells, but occurs in different ways in different species. For example, animals undergo an "open" mitosis, where the nuclear envelope breaks down before the chromosomes separate, while fungi such as *Aspergillus nidulans* and *Saccharomyces cerevisiae* (yeast) undergo a "closed" mitosis, where chromosomes divide within an intact cell nucleus. Prokaryotic cells, which lack a nucleus, divide by a process called binary fission.

The process of mitosis is complex and highly regulated. The sequence of events is divided into phases, corresponding to the completion of one set of activities and the start of the next. These stages are interphase, prophase, prometaphase, metaphase, anaphase and telophase. During the process of mitosis the pairs of chromosomes condense and attach to fibres that pull the sister chromatids to opposite sides of the cell. The cell then divides in cytokinesis, to produce two identical daughter cells.

Because cytokinesis usually occurs in conjunction with mitosis, "mitosis" is often used interchangeably with "mitotic phase". However, there are many cells where mitosis and cytokinesis occur separately, forming single cells with multiple nuclei. This occurs most notably among the fungi and slime moulds, but is found in various different groups. Even in animals, cytokinesis and mitosis may occur independently, for instance during certain stages of fruit fly embryonic development. Errors in mitosis can either kill a cell through apoptosis or cause mutations that may lead to cancer.

Overview

The primary result of mitosis is the division of the parent cell's genome into two daughter cells. The genome is composed of a number of chromosomes - complexes of tightly-coiled DNA that contain genetic information vital for proper cell function. Because each resultant daughter cell should be genetically identical to the parent cell, the parent cell must make a copy of each chromosome before mitosis. This occurs during S phase, in interphase, the period that precedes the mitotic phase in the cell cycle where preparation for mitosis occurs.

Each new chromosome now contains two identical copies of itself, called *sister chromatids*, attached together in a specialized region of the chromosome known as the *centromere*. Each sister chromatid is not considered a chromosome in itself, and a chromosome always contains two sister chromatids.

In most eukaryotes, the nuclear envelope that separates the DNA from the cytoplasm disassembles. The chromosomes align themselves in a line spanning the cell. Microtubules, essentially miniature strings, splay out from opposite ends of the cell and shorten, pulling apart the sister chromatids of each chromosome. As a matter of convention, each sister chromatid is now considered a chromosome, so they are renamed to *sister chromosomes*. As the cell elongates, corresponding sister chromosomes are pulled toward opposite ends. A new nuclear envelope forms around the separated sister chromosomes.

As mitosis completes cytokinesis is well underway. In animal cells, the cell pinches inward where the imaginary line used to be (the pinching of the cell membrane to form the two daughter cells is called the cleavage furrow), separating the two developing nuclei. In plant cells, the daughter cells will construct a new dividing cell wall between each other. Eventually, the mother cell will be split in half, giving rise to two daughter cells, each with an equivalent and complete copy of the original genome.

Prokaryotic cells undergo a process similar to mitosis called binary fission. However, prokaryotes cannot be properly said to undergo mitosis because they lack a nucleus and only have a single chromosome with no centromere.

Phases of Cell Cycle and Mitosis

Interphase

The mitotic phase is a relatively short period of the cell cycle. It alternates with the much longer *interphase*, where the cell prepares itself for cell division. Interphase is therefore not part of mitosis. Interphase is divided into three phases, G_1 (first gap), S (synthesis), and G_2 (second gap). During all three phases, the cell grows by producing proteins and cytoplasmic organelles. However, chromosomes are replicated only during the S phase. Thus, a cell grows (G_1), continues to grow as it duplicates its chromosomes (S), grows more and prepares for mitosis (G_2), and finally it divides (M) before restarting the cycle.

Preprophase

In plant cells only, prophase is preceded by a pre-prophase stage. In highly vacuolated plant cells, the nucleus has to migrate into the centre of the cell before mitosis can begin. This is achieved through the formation of a phragmosome, a transverse sheet of cytoplasm that bisects the cell along the future plane of cell division. In addition to phragmosome formation, preprophase is characterized by the formation of a ring of microtubules and actin filaments (called preprophase band) underneath the plasma membrane around the equatorial plane of the future mitotic spindle.

This band marks the position where the cell will eventually divide. The cells of higher plants (such as the flowering plants) lack centrioles: with microtubules forming a spindle on the surface of the nucleus and then being organized into a spindle by the chromosomes themselves, after the nuclear membrane breaks down. The preprophase band disappears during nuclear envelope disassembly and spindle formation in prometaphase.

Prophase

Normally, the genetic material in the nucleus is in a loosely bundled coil called chromatin. At the onset of prophase, chromatin condenses together into a highly ordered structure called a chromosome. Since the genetic material has already been duplicated earlier in S phase, the replicated chromosomes have two sister chromatids, bound together at the centromere by the cohesion complex. Chromosomes are visible at high magnification through a light microscope.

Close to the nucleus are structures called centrosomes, which are made of a pair of centrioles. The centrosome is the coordinating centre for the cell's microtubules. A cell inherits a single centrosome at cell division, which replicates before a new mitosis begins, giving a pair of centrosomes. The two centrosomes nucleate microtubules (which may be thought of as cellular ropes or poles) to form the spindle by polymerizing soluble tubulin. Molecular motor proteins then push the centrosomes along these microtubules to opposite sides of the cell. Although centrioles help organize microtubule assembly, they are not essential for the formation of the spindle, since they are absent from plants, and centrosomes are not always used in meiosis.

Prometaphase

The nuclear envelope disassembles and microtubules invade the nuclear space. This is called open mitosis, and it occurs in most multicellular organisms. Fungi and some protists, such as algae or trichomonads, undergo a variation called closed mitosis where the spindle forms inside the nucleus or its microtubules are able to penetrate an intact nuclear envelope.

Each chromosome forms two kinetochores at the centromere, one attached at each chromatid. A kinetochore is a complex protein structure that is analogous to a ring for the microtubule hook; it is the point where microtubules attach themselves to the chromosome. Although the kinetochore structure and function are not fully understood, it is known that it contains some form of molecular motor. When a microtubule connects with the kinetochore, the motor activates, using energy from ATP to "crawl" up the tube toward the originating centrosome. This motor activity, coupled with polymerisation and depolymerisation of microtubules, provides the pulling force necessary to later separate the chromosome's two chromatids.

When the spindle grows to sufficient length, *kinetochore microtubules* begin searching for kinetochores to attach to. A number of *nonkinetochore microtubules* find and interact with corresponding nonkinetochore microtubules from the opposite centrosome to form the mitotic spindle. Prometaphase is sometimes considered part of prophase.

In the fishing pole analogy, the kinetochore would be the "hook" that catches a sister chromatid or "fish". The centrosome acts as the "reel" that draws in the spindle fibres or "fishing line".

Metaphase

As microtubules find and attach to kinetochores in prometaphase, the centromeres of the chromosomes convene along the *metaphase plate* or *equatorial plane*, an imaginary line that is equidistant from the two centrosome poles. This even alignment is due to the counterbalance of the pulling powers generated by the opposing kinetochores, analogous to a tug-of-war between people of equal strength. In certain types of cells, chromosomes do not line up at the metaphase plate and instead move back and forth between the poles randomly, only roughly lining up along the midline. Metaphase comes from the Greek meaning "after."

Because proper chromosome separation requires that every kinetochore be attached to a bundle of microtubules (spindle fibres), it is thought that unattached kinetochores generate a signal to prevent premature progression to anaphase without all chromosomes being aligned. The signal creates the *mitotic spindle checkpoint.*

Anaphase

When every kinetochore is attached to a cluster of microtubules and the chromosomes have lined up along the metaphase plate, the cell proceeds to anaphase.

Two events then occur; First, the proteins that bind sister chromatids together are cleaved, allowing them to separate. These sister chromatids, which have now become distinct sister chromosomes, are pulled apart by shortening kinetochore microtubules and move toward the respective centrosomes to which they are attached. Next, the nonkinetochore microtubules elongate, pulling the centrosomes (and the set of chromosomes to which they are attached) apart to opposite ends of the cell. The force that causes the centrosomes to move towards the ends of the cell is still unknown, although there is a theory that suggests that the rapid assembly and breakdown of microtubules may cause this movement.

These two stages are sometimes called early and late anaphase. Early anaphase is usually defined as the separation of the sister chromatids, while late anaphase is the elongation of the microtubules and the chromosomes being pulled farther apart. At the end of anaphase, the cell has succeeded in separating identical copies of the genetic material into two distinct populations.

Telophase

Telophase is a reversal of prophase and prometaphase events. It "cleans up" the after effects of mitosis. At telophase, the nonkinetochore microtubules continue to lengthen, elongating the cell even more. Corresponding sister chromosomes attach at opposite ends of the cell. A new nuclear envelope, using fragments of the parent cell's nuclear membrane, forms around each set of separated sister chromosomes. Both sets of chromosomes, now surrounded by new nuclei, unfold back into chromatin. Mitosis is complete, but cell division is not yet complete.

Cytokinesis

Cytokinesis is often mistakenly thought to be the final part of telophase; however, cytokinesis is a separate process that begins at the same time as telophase. Cytokinesis is technically not even a phase of mitosis, but rather a separate process, necessary for completing cell division. In animal cells, a cleavage furrow (pinch) containing a contractile ring develops where the metaphase plate used to be, pinching off the separated nuclei. In both animal and plant cells, cell division is also driven by vesicles derived from the Golgi apparatus, which move along microtubules to the middle of the cell. In plants this structure coalesces into a cell plate at the centre of the phragmoplast and develops into a cell wall, separating the two nuclei. The phragmoplast is a microtubule structure typical for higher plants, whereas some green algae use a phycoplast microtubule array during cytokinesis. Each daughter cell has a complete copy of the genome of its parent cell. The end of cytokinesis marks the end of the M-phase.

Significance

Mitosis is important for the maintenance of the chromosomal set; each cell formed receives chromosomes that are alike in composition and equal in number to the chromosomes of the parent cell. Transcription is generally believed to cease during mitosis, but epigenetic mechanisms such as bookmarking function during this stage of the cell cycle to ensure that the "memory" of which genes were active prior to entry into mitosis are transmitted to the daughter cells.

Consequences of Errors

Although errors in mitosis are rare, the process may go wrong, especially during early cellular divisions in the zygote. Mitotic errors can be especially dangerous to the organism because future offspring from this parent cell will carry the same disorder. In *non-disjunction*, a chromosome may fail to separate during anaphase. One daughter cell will receive both sister chromosomes and the other will receive none. This results in the former cell having three chromosomes containing the same genes (two sisters and a homologue), a condition known as *trisomy*, and the latter cell having only one chromosome (the homologous chromosome), a condition known as *monosomy*. These cells are considered aneuploid, a condition often associated with cancer. Mitosis is a traumatic process.

The cell goes through dramatic changes in ultrastructure, its organelles disintegrate and reform in a matter of hours, and chromosomes are jostled constantly by probing microtubules. Occasionally, chromosomes may become damaged. An arm of the chromosome may be broken and the fragment lost, causing deletion. The fragment may incorrectly reattach to another, non-homologous chromosome, causing translocation. It may reattach to the original chromosome, but in reverse orientation, causing inversion. Or, it may be treated erroneously as a separate chromosome, causing chromosomal duplication. The effect of these genetic abnormalities depends on the specific nature of the error. It may range from no noticeable effect to cancer induction or organism death.

Endomitosis

Endomitosis is a variant of mitosis without nuclear or cellular division, resulting in cells with many copies of the same chromosome occupying a single nucleus. This process may also be referred to as endoreduplication and the cells as endoploid. An example of a cell that goes through endomitosis is the megakaryocyte.

Meiosis

In biology, meiosis is a process of reductional division in which the number of chromosomes per cell is cut in half. In animals, meiosis always results in the formation of gametes, while in other organisms it can give rise to spores. As with mitosis, before meiosis begins, the DNA in the original cell is replicated during S-phase of the cell cycle. Two cell divisions separate the replicated chromosomes into four haploid gametes or spores. Meiosis is essential for sexual reproduction and therefore occurs in all eukaryotes (including single-celled organisms) that reproduce sexually. A few eukaryotes, notably the Bdelloid rotifers, have lost the ability to carry out meiosis and have acquired the ability to reproduce by parthenogenesis. Meiosis does not occur in archaea or bacteria, which reproduce via asexual processes such as binary fission.

During meiosis, the genome of a diploid germ cell, which is composed of long segments of DNA packaged into chromosomes, undergoes DNA replication followed by two rounds of division, resulting in four haploid cells. Each of these cells contains one complete set of chromosomes, or half of the genetic content of the original cell. If meiosis produces gametes, these cells must fuse during fertilization to create a new diploid cell, or zygote before any new growth can occur. Thus, the division mechanism of meiosis is a reciprocal process to the joining of two genomes that occurs at fertilization. Because the chromosomes of each parent undergo homologous recombination during meiosis, each gamete, and thus each zygote, will have a unique genetic *blueprint* encoded in its DNA. Together, meiosis and fertilization constitute sexuality in the eukaryotes, and generate genetically distinct individuals in populations. In all plants, and in many protists, meiosis results in the formation of haploid cells that can divide vegetatively without undergoing fertilization, referred to as spores. In these groups, gametes are produced by mitosis. Meiosis uses many of the same biochemical mechanisms employed during mitosis to accomplish the redistribution of chromosomes. There are several features unique to meiosis, most importantly the pairing and recombination between homologous chromosomes. Meiosis comes from the root -meio, meaning less.

History

Meiosis was discovered and described for the first time in sea urchin eggs in 1876, by noted German biologist Oscar Hertwig (1849–1922). It was described again in 1883, at the level of chromosomes, by Belgian zoologist Edouard Van Beneden (1846–1910), in *Ascaris* worms' eggs. The significance of meiosis for reproduction and inheritance, however, was described only in 1890 by German biologist August Weismann (1834–1914), who noted that two cell divisions were necessary to transform one diploid cell into four haploid cells if the number of chromosomes had to be maintained. In 1911 the

American geneticist Thomas Hunt Morgan (1866–1945) observed crossover in *Drosophila melanogaster* meiosis and provided the first genetic evidence that genes are transmitted on chromosomes.

Evolution

Meiosis is thought to have appeared 1.4 billion years ago. The only supergroup of eukaryotes which does not have meiosis in all organisms is excavata. The other five major supergroups, opisthokonts, amoebozoa, rhizaria, archaeplastida and chromalveolates all seem to have genes for meiosis universally present, even if not always functional. Some excavata species do have meiosis which is consistent with the hypothesis that this group is an ancient, paraphyletic grade. An example of a eukaryotic organism in which meiosis does not exist is euglenoid.

Occurrence of Meiosis in Eukaryotic Life Cycles

Meiosis occurs in eukaryotic life cycles involving sexual reproduction, comprising of the constant cyclical process of meiosis and fertilization. This takes place alongside normal mitotic cell division. In multicellular organisms, there is an intermediary step between the diploid and haploid transition where the organism grows. The organism will then produce the germ cells that continue in the life cycle. The rest of the cells, called somatic cells, function within the organism and will die with it. Cycling meiosis and fertilization events produces a series of transitions back and forth between alternating haploid and diploid states.

The organism phase of the life cycle can occur either during the diploid state (*gametic* or *diploid* life cycle), during the haploid state (*zygotic* or *haploid* life cycle), or both (*sporic* or *haplodiploid* life cycle, in which there two distinct organism phases, one during the haploid state and the other during the diploid state). In this sense, there are three types of life cycles that utilize sexual reproduction, differentiated by the location of the organisms phase(s). In the *gametic life cycle*, of which humans are a part, the species is diploid, grown from a diploid cell called the zygote. The organism's diploid germ-line stem cells undergo meiosis to create haploid gametes (the spermatozoa for males and ova for females), which fertilize to form the zygote. The diploid zygote undergoes repeated cellular division by mitosis to grow into the organism. Mitosis is a related process to meiosis that creates two cells that are genetically identical to the parent cell. The general principle is that mitosis creates somatic cells and meiosis creates germ cells. In the *zygotic life cycle* the species is haploid instead, spawned by the proliferation and differentiation of a single haploid cell called the gamete. Two organisms of opposing gender contribute their haploid germ cells to form a diploid zygote. The zygote undergoes meiosis immediately, creating four haploid cells. These cells undergo mitosis to create the organism. Many fungi and many protozoa are members of the zygotic life cycle.

Finally, in the *sporic life cycle*, the living organism alternates between haploid and diploid states. Consequently, this cycle is also known as the alternation of generations. The diploid organism's germ-line cells undergo meiosis to produce spores. The spores proliferate by mitosis, growing into a haploid organism. The haploid organism's germ cells then combine with another haploid organism's cells, creating the zygote. The zygote undergoes repeated mitosis and differentiation to become the diploid organism again. The sporic life cycle can be considered a fusion of the gametic and zygotic life cycles.

Process

Because meiosis is a "one-way" process, it cannot be said to engage in a cell cycle as mitosis does. However, the preparatory steps that lead up to meiosis are identical in pattern and name to the interphase of the mitotic cell cycle. Interphase is divided into three phases:

- *Growth 1 (G_1) Phase:* This is a very active period, where the cell synthesizes its vast array of proteins, including the enzymes and structural proteins it will need for growth. In G_1 stage each of the chromosomes consists of a single (very long) molecule of DNA. In humans, at this point cells are 46 chromosomes, 2N, identical to somatic cells.

- *Synthesis (S) Phase:* The genetic material is replicated: each of its chromosomes duplicates, so that each of the 46 chromosomes forms a second identical sister chromatid. The cell is still considered diploid because it still contains the same number of centromeres. The identical sister chromatids have not yet condensed into the densely packaged chromosomes visible with the light microscope. This will take place during prophase I in meiosis.

- *Growth 2 (G_2) Phase:* G_2 phase is absent in Meiosis.

Interphase is followed by meiosis I and then meiosis II. Meiosis I consists of separating the pairs of homologous chromosome, each made up of two sister chromatids, into two cells. One entire haploid content of chromosomes is contained in each of the resulting daughter cells; the first meiotic division therefore reduces the ploidy of the original cell by a factor of 2. Meiosis II consists of decoupling each chromosome's sister strands (chromatids), and segregating the individual chromatids into haploid daughter cells.

The two cells resulting from meiosis I divide during meiosis II, creating 4 haploid daughter cells. Meiosis I and II are each divided into prophase, metaphase, anaphase, and telophase stages, similar in purpose to their analogous subphases in the mitotic cell cycle. Therefore, meiosis includes the stages of meiosis I (prophase I, metaphase I, anaphase I, telophase I), and meiosis II (prophase II, metaphase II, anaphase II, telophase II). Meiosis generates genetic diversity in two ways: (1) independent alignment and subsequent separation of homologous chromosome pairs during the first meiotic division allows a random and independent selection of each chromosome segregates into each gamete; and (2) physical exchange of homologous chromosomal regions by homologous recombination during prophase I results in new combinations of DNA within chromosomes.

Phases of Meiosis

Meiosis takes place in several stages.

Meiosis I

Meiosis I separates homologous chromosomes, producing two haploid cells (N chromosomes, 23 in humans), so meiosis I is referred to as a reductional division. A regular diploid human cell contains 46 chromosomes and is considered 2N because it contains 23 pairs of homologous chromosomes. However, after meiosis I, although the cell contains 46 chromatids, it is only considered as being N, with 23 chromosomes. This is because later, in Anaphase I, the sister chromatids will remain together as the spindle fibres pulls the pair toward the pole of the new cell. In meiosis II, an equational division similar to mitosis will occur whereby the sister chromatids are finally split, creating a total of 4 haploid cells (23 chromosomes, N) per daughter cell from the first division.

Prophase I

During prophase I, DNA is exchanged between homologous chromosomes in a process called homologous recombination. This often results in chromosomal crossover. The new combinations of DNA created during crossover are a significant source of genetic variation, and may result in beneficial new combinations of alleles. The paired and replicated chromosomes are called bivalents or tetrads, which have two chromosomes and four chromatids, with one chromosome coming from each parent. At this stage, non-sister chromatids may cross-over at points called chiasmata (plural; singular chiasma).

Leptotene

The first stage of prophase I is the *leptotene* stage, also known as *leptonema*, from Greek words meaning "thin threads". During this stage, individual chromosomes begin to condense into long strands within the nucleus. However the two sister chromatids are still so tightly bound that they are indistinguishable from one another.

Zygotene

The *zygotene* stage, also known as *zygonema*, from Greek words meaning "paired threads", occurs as the chromosomes approximately line up with each other into homologous chromosome pairs. This is called the bouquet stage because of the way the telomeres cluster at one end of the nucleus. At this stage, the synapsis (pairing/coming together) of homologous chromosomes takes place.

Pachytene

The *pachytene* stage, also known as *pachynema*, from Greek words meaning "thick threads", contains the following chromosomal crossover. Nonsister chromatids of homologous chromosomes randomly exchange segments of genetic information over regions of homology. Sex chromosomes, however, are not wholly identical, and only exchange information over a small region of homology. Exchange takes place at sites where *recombination nodules* (the chiasmata) have formed. The exchange of information between the non-sister chromatids results in a recombination of information; each chromosome has the complete set of information it had before, and there are no gaps formed as a result of the process. Because the chromosomes cannot be distinguished in the synaptonemal complex, the actual act of crossing over is not perceivable through the microscope.

Diplotene

During the *diplotene* stage, also known as *diplonema*, from Greek words meaning "two threads", the synaptonemal complex degrades and homologous chromosomes separate from one another a little. The chromosomes themselves uncoil a bit, allowing some transcription of DNA. However, the homologous chromosomes of each bivalent remain tightly bound at chiasmata, the regions where crossing-over occurred. The chiasmata remain on the chromosomes until they are severed in Anaphase I. In human fetal oogenesis all developing oocytes develop to this stage and stop before birth. This suspended state is referred to as the *dictyotene stage* and remains so until puberty. In males, only spermatogonia (spermatogenesis) exist until meiosis begins at puberty.

Diakinesis

Chromosomes condense further during the *diakinesis* stage, from Greek words meaning "moving through". This is the first point in meiosis where the four parts of the tetrads are actually visible. Sites of crossing over entangle together, effectively overlapping, making chiasmata clearly visible. Other than this observation, the rest of the stage closely resembles prometaphase of mitosis; the nucleoli disappear, the nuclear membrane disintegrates into vesicles, and the meiotic spindle begins to form.

Synchronous Processes

During these stages, two centrosomes, containing a pair of centrioles in animal cells, migrate to the two poles of the cell. These centrosomes, which were duplicated during S-phase, function as microtubule organizing centres nucleating microtubules, which are essentially cellular ropes and poles. The microtubules invade the nuclear region after the nuclear envelope disintegrates, attaching to the chromosomes at the kinetochore. The kinetochore functions as a motor, pulling the chromosome along the attached microtubule toward the originating centriole, like a train on a track. There are four kinetochores on each tetrad, but the pair of kinetochores on each sister chromatid fuses and functions as a unit during meiosis I. Microtubules that attach to the kinetochores are known as *kinetochore microtubules*. Other microtubules will interact with microtubules from the opposite centriole: these are called *nonkinetochore microtubules* or *polar microtubules*. A third type of microtubules, the aster microtubules, radiates from the centrosome into the cytoplasm or contacts components of the membrane skeleton.

Metaphase I

Homologous pairs move together along the metaphase plate: As kinetochore microtubules from both centrioles attach to their respective kinetochores, the homologous chromosomes align along an equatorial plane that bisects the spindle, due to continuous counterbalancing forces exerted on the bivalents by the microtubules emanating from the two kinetochores of homologous chromosomes. The physical basis of the independent assortment of chromosomes is the random orientation of each bivalent along the metaphase plate, with respect to the orientation of the other bivalents along the same equatorial line.

Anaphase I

Kinetochore (bipolar spindles) microtubules shorten, severing the recombination nodules and pulling homologous chromosomes apart. Since each chromosome has only one functional unit of a pair of kinetochores, whole chromosomes are pulled toward opposing poles, forming two haploid sets. Each chromosome still contains a pair of sister chromatids. Nonkinetochore microtubules lengthen, pushing the centrioles farther apart. The cell elongates in preparation for division down the centre.

Telophase I

The last meiotic division effectively ends when the chromosomes arrive at the poles. Each daughter cell now has half the number of chromosomes but each chromosome consists of a pair of chromatids. The microtubules that make up the spindle network disappear, and a new nuclear membrane surrounds each haploid set. The chromosomes uncoil back into chromatin. Cytokinesis, the pinching of the cell membrane in animal cells or the formation of the cell wall in plant cells, occurs, completing the creation of two daughter cells. Sister chromatids remain attached during telophase I.

Cells may enter a period of rest known as interkinesis or interphase II. No DNA replication occurs during this stage.

Meiosis II

Meiosis II is the second part of the meiotic process. Much of the process is similar to mitosis. The end result is production of four haploid cells (23 chromosomes, 1N in humans) from the two haploid cells (23 chromosomes, 1N * each of the chromosomes consisting of two sister chromatids) produced in meiosis I. The four main steps of Meiosis II are: Prophase II, Metaphase II, Anaphase II, and Telophase II. In prophase II we see the disappearance of the nucleoli and the nuclear envelope again as well as the shortening and thickening of the chromatids. Centrioles move to the polar regions and arrange spindle fibres for the second meiotic division.

In metaphase II, the centromeres contain two kinetochores that attach to spindle fibres from the centrosomes (centrioles) at each pole. The new equatorial metaphase plate is rotated by 90 degrees when compared to meiosis I, perpendicular to the previous plate.

This is followed by anaphase II, where the centromeres are cleaved, allowing microtubules attached to the kinetochores to pull the sister chromatids apart. The sister chromatids by convention are now called sister chromosomes as they move toward opposing poles. The process ends with telophase II, which is similar to telophase I, and is marked by uncoiling and lengthening of the chromosomes and the disappearance of the spindle.

Nuclear envelopes reform and cleavage or cell wall formation eventually produces a total of four daughter cells, each with a haploid set of chromosomes. Meiosis is now complete and ends up with four new daughter cells.

Significance

Meiosis facilitates stable sexual reproduction. Without the halving of ploidy, or chromosome count, fertilization would result in zygotes that have twice the number of chromosomes as the zygotes from the previous generation. Successive generations would have an exponential increase in chromosome count.

In organisms that are normally diploid, polyploidy, the state of having three or more sets of chromosomes, results in extreme developmental abnormalities or lethality. Polyploidy is poorly tolerated in most animal species. Plants, however, regularly produce fertile, viable polyploids. Polyploidy has been implicated as an important mechanism in plant speciation.

Most importantly, recombination and independent assortment of homologous chromosomes allow for a greater diversity of genotypes in the population. This produces genetic variation in gametes that promote genetic and phenotypic variation in a population of offspring.

Nondisjunction

The normal separation of chromosomes in meiosis I or sister chromatids in meiosis II is termed *disjunction*. When the separation is not normal, it is called nondisjunction. This results in the production of gametes which have either too many or too few of a particular chromosome, and is a common mechanism for trisomy or monosomy. Nondisjunction can occur in the meiosis I or meiosis II, phases of cellular reproduction, or during mitosis.

This is a cause of several medical conditions in humans (such as):

- Down Syndrome - trisomy of chromosome 21
- Patau Syndrome - trisomy of chromosome 13
- Edward Syndrome - trisomy of chromosome 18
- Klinefelter Syndrome - extra X chromosomes in males - i.e. XXY, XXXY, XXXXY
- Turner Syndrome - lacking of one X chromosome in females - i.e. XO
- Triple X syndrome - an extra X chromosome in females
- XYY Syndrome - an extra Y chromosome in males.

Meiosis in Mammals

In females, meiosis occurs in cells known as oogonia (singular: oogonium). Each oogonium that initiates meiosis will divide twice to form a single oocyte and two polar bodies. However, before these divisions occur, these cells stop at the diplotene stage of meiosis I and lie dormant within a protective shell of somatic cells called the follicle. Follicles begin growth at a steady pace in a process known as folliculogenesis, and a small number enter the menstrual cycle. Menstruated oocytes continue meiosis I and arrest at meiosis II until fertilization. The process of meiosis in females occurs during oogenesis, and differs from the typical meiosis in that it features a long period of meiotic arrest known as the Dictyate stage and lacks the assistance of centrosomes.

In males, meiosis occurs in precursor cells known as spermatogonia that divide twice to become sperm. These cells continuously divide without arrest in the seminiferous tubules of the testicles. Sperm is produced at a steady pace. The process of meiosis in males occurs during spermatogenesis. In female mammals, meiosis begins immediately after primordial germ cells migrate to the ovary in the embryo, but in the males, meiosis begins years later at the time of puberty. It is retinoic acid, derived from the primitive kidney (mesonephros) that stimulates meiosis in ovarian oogonia.

Tissues of the male testis suppress meiosis by degrading retinoic acid, a stimulator of meiosis. This is overcome at puberty when cells within seminiferous tubules called Sertoli cells start making their own retinoic acid. Sensitivity to retinoic acid is also adjusted by proteins called nanos and DAZL. Meiosis involves Spermatocytes.

Chromosomes in Prokaryotes

The prokaryotes – bacteria and archaea – typically have a single circular chromosome, but many variations do exist. Most bacteria have a single circular chromosome that can range in size from only 160,000 base pairs in the endosymbiotic bacterium *Candidatus Carsonella ruddii*, to 12,200,000 base pairs in the soil-dwelling bacterium *Sorangium cellulosum*. Spirochaetes of the genus *Borrelia* are a notable exception to this arrangement, with bacteria such as *Borrelia burgdorferi*, the cause of Lyme disease, containing a single linear chromosome.

Structure in Sequences

Prokaryotic chromosomes have less sequence-based structure than eukaryotes. Bacteria typically have a single point (the origin of replication) from which replication starts, whereas some archaea contain multiple replication origins. The genes in prokaryotes are often organized in operons, and do not usually contain introns, unlike eukaryotes.

DNA Packaging

Prokaryotes do not possess nuclei. Instead, their DNA is organized into a structure called the nucleoid. The nucleoid is a distinct structure and occupies a defined region of the bacterial cell. This structure is, however, dynamic and is maintained and remodeled by the actions of a range of histone-like proteins, which associate with the bacterial chromosome. In archaea, the DNA in chromosomes is even more organized, with the DNA packaged within structures similar to eukaryotic nucleosomes.

Bacterial chromosomes tend to be tethered to the plasma membrane of the bacteria. In molecular biology application, this allows for its isolation from plasmid DNA by centrifugation of lysed bacteria and pelleting of the membranes (and the attached DNA). Prokaryotic chromosomes and plasmids are, like eukaryotic DNA, generally supercoiled. The DNA must first be released into its relaxed state for access for transcription, regulation, and replication. "beads-on-a-string" fibre, the 30nm fibre and the metaphase chromosome.

Number of Chromosomes in Various Organisms

Eukaryotes

These tables give the total number of chromosomes (including sex chromosomes) in a cell nucleus. For example, human cells are diploid and have 22 different types of autosome, each present as two copies, and two sex chromosomes. This gives 46 chromosomes in total. Other organisms have more than two copies of their chromosomes, such as bread wheat, which is *hexaploid* and has six copies of seven different chromosomes – 42 chromosomes in total. Normal members of a particular eukaryotic species all have the same number of nuclear chromosomes. Other eukaryotic chromosomes, i.e., mitochondrial and plasmid-like small chromosomes, are much more variable in number, and there may be thousands of copies per cell.

Asexually reproducing species have one set of chromosomes, which are the same in all body cells. However, asexual species can be either haploid or diploid.

Sexually reproducing species have somatic cells (body cells), which are diploid [2n] having two sets of chromosomes, one from the mother and one from the father. Gametes, reproductive cells, are haploid [n]: They have one set of chromosomes. Gametes are produced by meiosis of a diploid germ line cell. During meiosis, the matching chromosomes of father and mother can exchange small parts of themselves (crossover), and thus create new chromosomes that are not inherited solely from either parent. When a male and a female gamete merge (fertilization), a new diploid organism is formed. Some animal and plant species are polyploid [Xn]: They have more than two sets of homologous chromosomes. Plants important in agriculture such as tobacco or wheat are often polyploid, compared to their ancestral species. Wheat has a haploid number of seven chromosomes, still seen in some cultivars as well as the wild progenitors. The more-common pasta and bread wheats are polyploid, having 28 (tetraploid) and 42 (hexaploid) chromosomes, compared to the 14 (diploid) chromosomes in the wild wheat.

Prokaryotes

Prokaryote species generally have one copy of each major chromosome, but most cells can easily survive with multiple copies. For example, *Buchnera*, a symbiont of aphids has multiple copies of its chromosome, ranging from 10–400 copies per cell. However, in some large bacteria, such as *Epulopiscium fishelsoni* up to 100,000 copies of the chromosome can be present. Plasmids and plasmid-like small chromosomes are, as in eukaryotes, very variable in copy number. The number of plasmids in the cell is almost entirely determined by the rate of division of the plasmid – fast division causes high copy number, and vice versa.

Karyotype

In general, the karyotype is the characteristic chromosome complement of a eukaryote species. The preparation and study of karyotypes is part of cytogenetics.

Although the replication and transcription of DNA is highly standardized in eukaryotes, *the same cannot be said for their karyotypes*, which are often highly variable. There may be variation between species in chromosome number and in detailed organization. In some cases, there is significant variation within species. Often there is:

1. variation between the two sexes
2. variation between the germ-line and soma (between gametes and the rest of the body)
3. variation between members of a population, due to balanced genetic polymorphism
4. geographical variation between races
5. mosaics or otherwise abnormal individuals.

Also, variation in karyotype may occur during development from the fertilised egg.

The technique of determining the karyotype is usually called *karyotyping*. Cells can be locked part-way through division (in metaphase) in vitro (in a reaction vial) with colchicine. These cells are then stained, photographed, and arranged into a *karyogram*, with the set of chromosomes arranged, autosomes in order of length, and sex chromosomes (here X/Y) at the end:

Like many sexually reproducing species, humans have special gonosomes (sex chromosomes, in contrast to autosomes). These are XX in females and XY in males.

Historical Note

Investigation into the human karyotype took many years to settle the most basic question. How many chromosomes does a normal diploid human cell contain? In 1912, Hans von Winiwarter reported 47 chromosomes in spermatogonia and 48 in oogonia, concluding an XX/XO sex determination mechanism. Painter in 1922 was not certain whether the diploid number of man is 46 or 48, at first favouring 46. He revised his opinion later from 46 to 48, and he correctly insisted on man's having an XX/XY system. New techniques were needed to definitively solve the problem:

1. Using cells in culture
2. Pretreating cells in a hypotonic solution, which swells them and spreads the chromosomes
3. Arresting mitosis in metaphase by a solution of colchicine
4. Squashing the preparation on the slide forcing the chromosomes into a single plane
5. Cutting up a photomicrograph and arranging the result into an indisputable karyogram.

It took until the mid-1950s for it to become generally accepted that the human karyotype include only 46 chromosomes. Considering the techniques of Winiwarter and Painter, their results were quite remarkable. Chimpanzees (the closest living relatives to modern humans) have 48 chromosomes.

Chromosomal Aberrations

Chromosomal aberrations are disruptions in the normal chromosomal content of a cell, and are a major cause of genetic conditions in humans, such as Down syndrome. Some chromosome abnormalities do not cause disease in carriers, such as translocations, or chromosomal inversions, although they may lead to a higher chance of birthing a child with a chromosome disorder. Abnormal numbers of chromosomes or chromosome sets, aneuploidy, may be lethal or give rise to genetic disorders. Genetic counseling is offered for families that may carry a chromosome rearrangement.

The gain or loss of DNA from chromosomes can lead to a variety of genetic disorders. Human examples include:

- Cri du chat, which is caused by the deletion of part of the short arm of chromosome 5. "Cri du chat" means "cry of the cat" in French, and the condition was so-named because affected babies make high-pitched cries that sound like those of a cat. Affected individuals have wide-set eyes, a small head and jaw, moderate to severe mental health issues, and are very short.

- Down syndrome, usually is caused by an extra copy of chromosome 21 (trisomy 21). Characteristics include decreased muscle tone, stockier build, asymmetrical skull, slanting eyes and mild to moderate developmental disability.

- Edwards syndrome, which is the second-most-common trisomy; Down syndrome is the most common. It is a trisomy of chromosome 18. Symptoms include motor retardation, developmental disability and numerous congenital anomalies causing serious health problems. Ninety percent die in infancy; however, those that live past their first birthday usually are quite healthy thereafter. They have a characteristic clenched hands and overlapping fingers.

- Idic15, abbreviation for Isodicentric 15 on chromosome 15; also called the following names due to various researches, but they all mean the same; IDIC(15), Inverted duplication 15, extra Marker, Inv dup 15, partial tetrasomy 15

- Jacobsen syndrome, also called the terminal 11q deletion disorder. This is a very rare disorder. Those affected have normal intelligence or mild developmental disability, with poor expressive language skills. Most have a bleeding disorder called Paris-Trousseau syndrome.

- Klinefelter's syndrome (XXY). Men with Klinefelter syndrome are usually sterile, and tend to have longer arms and legs and to be taller than their peers. Boys with the syndrome are often shy and quiet, and have a higher incidence of speech delay and dyslexia. During puberty, without testosterone treatment, some of them may develop gynecomastia.

- Patau Syndrome, also called D-Syndrome or trisomy-13. Symptoms are somewhat similar to those of trisomy-18, but they do not have the characteristic hand shape.

- Small supernumerary marker chromosome. This means there is an extra, abnormal chromosome. Features depend on the origin of the extra genetic material. Cat-eye syndrome and isodicentric chromosome 15 syndrome (or Idic15) are both caused by a supernumerary marker chromosome, as is Pallister-Killian syndrome.

- Triple-X syndrome (XXX). XXX girls tend to be tall and thin. They have a higher incidence of dyslexia.

- Turner syndrome (X instead of XX or XY). In Turner syndrome, female sexual characteristics are present but underdeveloped. People with Turner syndrome often have a short stature, low hairline, abnormal eye features and bone development and a "caved-in" appearance to the chest.

- XYY syndrome. XYY boys are usually taller than their siblings. Like XXY boys and XXX girls, they are somewhat more likely to have learning difficulties.

- Wolf-Hirschhorn syndrome, which is caused by partial deletion of the short arm of chromosome 4. It is characterized by severe growth retardation and severe to profound mental health issues.

Chromosomal mutations produce changes in whole chromosomes (more than one gene) or in the number of chromosomes present.

- Deletion – loss of part of a chromosome
- Duplication – extra copies of a part of a chromosome
- Inversion – reverse the direction of a part of a chromosome
- Translocation – part of a chromosome breaks off and attaches to another chromosome.

Most mutations are neutral – have little or no effect. Chromosomal aberrations are the changes in the structure of chromosomes. It has a great role in evolution. A detailed graphical display of all human chromosomes and the diseases annotated at the correct spot may be found at the Oak Ridge National Laboratory.

Human Chromosomes

Chromosomes can be divided into two types—autosomes, and sex chromosomes. Certain genetic traits are linked to your sex, and are passed on through the sex chromosomes. The autosomes contain the rest of the genetic hereditary information. All act in the same way during cell division. Human cells have 23 pairs of large linear nuclear chromosomes, (22 pairs of autosomes and one pair of sex chromosomes) giving a total of 46 per cell. In addition to these, human cells have many hundreds of copies of the mitochondrial genome. Sequencing of the human genome has provided a great deal of information about each of the chromosomes. Below is a table compiling statistics for the chromosomes, based on the Sanger Institute's human genome information in the Vertebrate Genome Annotation (VEGA) database. Number of genes is an estimate as it is in part based on gene predictions. Total chromosome length is an estimate as well, based on the estimated size of unsequenced heterochromatin regions.

Chromosome	Genes	Total bases	Sequenced bases
1	4,220	247,199,719	224,999,719
2	1,491	242,751,149	237,712,649
3	1,550	199,446,827	194,704,827
4	446	191,263,063	187,297,063
5	609	180,837,866	177,702,766

Contd...

Chromosome	Genes	Total bases	Sequenced bases
6	2,281	170,896,993	167,273,993
7	2,135	158,821,424	154,952,424
8	1,106	146,274,826	142,612,826
9	1,920	140,442,298	120,312,298
10	1,793	135,374,737	131,624,737
11	379	134,452,384	131,130,853
12	1,430	132,289,534	130,303,534
13	924	114,127,980	95,559,980
14	1,347	106,360,585	88,290,585
15	921	100,338,915	81,341,915
16	909	88,822,254	78,884,754
17	1,672	78,654,742	77,800,220
18	519	76,117,153	74,656,155
19	1,555	63,806,651	55,785,651
20	1,008	62,435,965	59,505,254
21	578	46,944,323	34,171,998
22	1,092	49,528,953	34,893,953
X (sex chromosome)	1,846	154,913,754	151,058,754
Y (sex chromosome)	454	57,741,652	25,121,652
Total	32,185	3,079,843,747	2,857,698,560

8

Genetic Disorders

A genetic disorder is an illness caused by abnormalities in genes or chromosomes. While some diseases, such as cancer, are due in part to genetic disorders, they can also be caused by environmental factors. Most disorders are quite rare and affect one person in every several thousands or millions. Some types of recessive gene disorders confer an advantage in the heterozygous state in certain environments.

Single Gene Disorder

A single gene disorder is the result of a single mutated gene. There are estimated to be over 4000 human diseases caused by single gene defects. Single gene disorders can be passed on to subsequent generations in several ways. Genomic imprinting and uniparental disomy, however, may affect inheritance patterns. The divisions between recessive and dominant types are not "hard and fast" although the divisions between autosomal and X-linked types are (since the latter types are distinguished purely based on the chromosomal location of the gene).

For example, achondroplasia is typically considered a dominant disorder, but children with two genes for achondroplasia have a severe skeletal disorder that achondroplasics could be viewed as carriers of. Sickle-cell anemia is also considered a recessive condition, but heterozygous carriers have increased resistance to malaria in early childhood, which could be described as a related dominant condition. When a couple where one partner or both are sufferers or carriers of a single gene disorder and wish to have a child they can do so through IVF whichs means they can then have PGD (pre-implantation genetic diagnosis) to check whether the fertilised egg has had the genetic disorder passed on. Autosomal Dominant.

Only one mutated copy of the gene will be necessary for a person to be affected by an autosomal dominant disorder. Each affected person usually has one affected parent. There is a 50% chance that a child will inherit the mutated gene. Conditions that are autosomal dominant often have low penetrance, which means that although only one mutated copy is needed, a relatively small proportion of those who inherit that mutation go on to develop the disease. Examples of this type of disorder are Huntington's disease, neurofibromatosis type 1, Marfan syndrome, hereditary nonpolyposis colorectal cancer, and hereditary multiple exostoses, which is a highly penetrant autosomal dominant disorder. Birth defects are also called congenital anomalies.

Autosomal Recessive

Two copies of the gene must be mutated for a person to be affected by an autosomal recessive disorder. An affected person usually has unaffected parents who each carry a single copy of the mutated gene (and are referred to as carriers). Two unaffected people who each carry one copy of the mutated gene have a 25% chance with each pregnancy of having a child affected by the disorder. Examples of this type of disorder are cystic fibrosis, sickle-cell disease (also partial sickle-cell disease), Tay-Sachs disease, Niemann-Pick disease, spinal muscular atrophy, Roberts syndrome, and Dry (otherwise known as "rice-brand") earwax.

X-linked Dominant

X-linked dominant disorders are caused by mutations in genes on the X chromosome. Only a few disorders have this inheritance pattern, with a prime example being X-linked hypophosphatemic rickets. Males and females are both affected in these disorders, with males typically being more severely affected than females. Some X-linked dominant conditions such as Rett syndrome, incontinentia pigmenti type 2 and Aicardi syndrome are usually fatal in males either in utero or shortly after birth, and are therefore predominantly seen in females. Exceptions to this finding are extremely rare cases in which boys with Klinefelter syndrome (47,XXY) also inherit an X-linked dominant condition and exhibit symptoms more similar to those of a female in terms of disease severity. The chance of passing on an X-linked dominant disorder differs between men and women. The sons of a man with an X-linked dominant disorder will all be unaffected (since they receive their father's Y chromosome), and his daughters will all inherit the condition. A woman with an X-linked dominant disorder has a 50% chance of having an affected fetus with each pregnancy, although it should be noted that in cases such as incontinentia pigmenti only female offspring are generally viable. In addition, although these conditions do not alter fertility per se, individuals with Rett syndrome or Aicardi syndrome rarely reproduce.

X-linked Recessive

X-linked recessive conditions are also caused by mutations in genes on the X chromosome. Males are more frequently affected than females, and the chance of passing on the disorder differs between men and women. The sons of a man with an X-linked recessive disorder will not be affected, and his daughters will carry one copy of the mutated gene. A woman who is a carrier of an X-linked recessive disorder (XX) has a 50% chance of having sons who are affected and a 50% chance of having daughters who carry one copy of the mutated gene and are therefore carriers. X-linked recessive conditions include the serious diseases Hemophilia A, Duchenne muscular dystrophy, and Lesch-Nyhan syndrome as well as common and less serious conditions such as male pattern baldness and red-green color blindness. X-linked recessive conditions can sometimes manifest in females due to skewed X-inactivation or monosomy X (Turner syndrome).

Y-linked

Y-linked disorders are caused by mutations on the Y chromosome. Because males inherit a Y chromosome from their fathers, *every* son of an affected father will be affected. Because females inherit an X chromosome from their fathers, female offspring of affected fathers are *never* affected.

Since the Y chromosome is relatively small and contains very few genes, there are relatively few Y-linked disorders. Often the symptoms include infertility, which may be circumvented with the help of some fertility treatments. Examples are male infertility and hypertrichosis pinnae.

Mitochondrial

This type of inheritance, also known as maternal inheritance, applies to genes in mitochondrial DNA. Because only egg cells contribute mitochondria to the developing embryo, only mothers can pass on mitochondrial conditions to their children. An example of this type of disorder is Leber's hereditary optic neuropathy. underwear

Multifactorial and Polygenic (Complex) Disorders

Genetic disorders may also be complex, multifactorial, or polygenic, meaning that they are likely associated with the effects of multiple genes in combination with lifestyle and environmental factors. Multifactorial disorders include heart disease and diabetes. Although complex disorders often cluster in families, they do not have a clear-cut pattern of inheritance. This makes it difficult to determine a person's risk of inheriting or passing on these disorders. Complex disorders are also difficult to study and treat because the specific factors that cause most of these disorders have not yet been identified.

On a pedigree, polygenic diseases do tend to "run in families", but the inheritance does not fit simple patterns as with Mendelian diseases. But this does not mean that the genes cannot eventually be located and studied. There is also a strong environmental component to many of them (e.g., blood pressure).

- asthma
- autoimmune diseases such as multiple sclerosis
- cancers
- ciliopathies
- cleft palate
- diabetes
- heart disease
- hypertension
- inflammatory bowel disease
- mental retardation
- mood disorder
- obesity
- refractive error.

Prognosis and Treatment of Genetic Disorders

Genetic disorders rarely have effective treatments, though gene therapy is being tested as a possible treatment for some genetic diseases, including some forms of retinitis pigmentosa

- Gauchers disease is a genetic disease affecting metabolism. It is more treatable then most other genetic diseases, and can be treated with enzyme replacement therapy, medication miglustat, and bone marrow transplantion.

Hypothyroidism

What is Hypothyroidism?

The thyroid is the largest endocrine gland in the body. It sits just below the larynx (voice box) and wraps around the trachea (windpipe). The thyroid gland produces thyroid hormone, which helps the body grow and develop. It also plays an important role in the body's metabolism (the processes in the body that use energy, such as eating, breathing, and regulating heat).

Hypothyroidism (or underactive thyroid) is a common condition in which the thyroid gland produces too little thyroid hormone. About 1 in 5,000 babies is born with congenital hypothyroidism, in which the thyroid fails to grow normally and cannot produce enough of its hormone. There is no known cause for most cases of congenital hypothyroidism. But about 10 to 20 percent of the time, the condition is caused by an inherited defect that alters the production of thyroid hormone.

The most common inherited form of hypothyroidism is a defect of the TPO (thyroid peroxidase) gene on chromosome 2. This gene plays an important role in thyroid hormone production.

How do People get Hypothyroidism?

Hypothyroidism may be caused by 1) an autoimmune disease that attacks the thyroid gland, 2) surgery or radiation to treat thyroid cancer and other conditions, or 3) rare and random genetic events in which a mutation is acquired during early development.

What are the Symptoms of Hypothyroidism?

In babies with the inherited form of hypothyroidism, the condition affects growth and cognitive development. It may cause mental retardation, delayed puberty, stunted growth, and ataxia (inability to coordinate muscle movements).

In adults, hypothyroidism slows the body's metabolism, making the patient feel mentally and physically sluggish. Symptoms may include weakness, fatigue, muscle aches, mood swings, hair loss, memory loss, or slow speech. A person's symptoms will depend upon how little thyroid hormone they produce, and for how long they have had the disorder.

When the body is deprived of thyroid hormone, the pituitary gland works overtime, producing extra thyroid-stimulating hormone

(TSH). This glut of TSH may enlarge the thyroid into a condition called a goiter.

How do Doctors Diagnose Hypothyroidism?

Babies are normally screened for hypothyroidism 24 hours after birth. A tiny sample of blood taken from the baby's heel is tested for low thyroid hormone levels or high thyroid-stimulating hormone (TSH) levels.

How is Hypothyroidism Treated?

Hormone replacement therapy: people with hypothyroidism must take a synthetic form of thyroid hormone every day to reduce their symptoms.

Interesting Facts about Hypothyroidism

One in 5,000 people has hypothyroidism.

Colon Cancer

What is Colon Cancer?

Cells normally grow and divide only when they are needed to keep our bodies functioning properly. But sometimes, the mechanisms that regulate cell growth stop working and cells divide out of control to form tumors. This is called cancer. When cancer develops in the cells lining the colon (the first part of the large intestine), it is called colon cancer.

People who have a history of colon cancer in their family are at greater risk of getting the disease themselves. The risk increases when a relative has had the disease before age 50. These families are considered high-risk, because they may have inherited one of two rare genetic conditions: FAP (familial adenomatous polyposis) or HNPCC (hereditary non-polyposis colon cancer). FAP is caused by mutations of the APC (adenomatous polyposis coli) gene on chromosome 5. APC is a tumour suppressor gene, which means that it prevents uncontrolled cell growth. People who inherit a mutated form of this gene develop growths called polyps in their colon. By age 15, they may have hundreds of these polyps. Polyps are not cancerous at first, but if they aren't treated, they will develop into colon cancer.

HNPCC (also called Lynch syndrome) is caused by mutations in one of several genes that fix damaged DNA. People who inherit one of these mutations have a much greater risk of accumulating mutations that will lead to uncontrolled cell growth and cancer.

How do People get Colon Cancer?

FAP and HNPCC are both inherited in an autosomal dominant pattern. If a parent has FAP or HNPCC, his or her children run a 50 percent risk of inheriting the mutated gene. Usually when a person inherits a defective gene it does not necessarily mean he or she will develop a malignant cancer. However, the APC gene strikingly predisposes one to colon cancer. People who inherit one bad copy of the APC gene are practically guaranteed to develop colon cancer by age 40. Similarly, people who inherit one bad copy of a gene associated with HNPCC have an 80 percent chance of getting colon cancer before. HNPCC also increases a person's risk of developing other cancers, including ovarian, stomach, brain, and liver.

What are the Symptoms of Colon Cancer?

Colon cancer affects the stomach and bowels. Common symptoms include: diarrhea or constipation, blood in the stool, vomiting, bloating, cramps, and unexplained weight loss.

How do Doctors Diagnose Colon Cancer?

When a patient shows symptoms of colon cancer, his or her doctor can screen for the disease using one of several tests:

- Fecal Occult Blood Test (FOBT) - Colon cancer can sometimes cause tiny dots of blood, too small for the eye to see, in the feces The FOBT test uses a special chemical to check the patient's stool sample for these traces of blood.
- Flexible-Sigmoidoscopy - Using a thin flexible tube called a simoidoscope, the doctor looks inside the patient's colon for growths called polyps.
- Double Contrast Barium Enema (DCBA) - A silvery-white metallic substance called barium is inserted up the patient's colon through the rectum. The barium outlines the patient's colon on an x-ray screen.
- Colonoscopy - Using a thin instrument called a colonoscope, the doctor looks inside the patient's colon. During the procedure, the doctor removes pieces of tissue (called a biopsy) to test them for cancer. If the doctor finds any polyps, he or she can also remove them. A newer method, called virtual colonoscopy, looks at the colon without going into the body, with an MRI or CT scan.

- DNA-Based stool test - This test examines DNA taken from a patient's stool sample to look for genetic defects associated with colon cancer.

How is Colon Cancer Treated?

Colon cancer is very treatable. In fact, about 90 percent of patients survive the disease after treatment. First, doctors stage the disease to see how far it has progressed. If the cancer has not spread to other tissues of the body, it can be treated with: special chemicals (chemotherapy) or radiation (powerful x-rays) that kill all rapidly dividing cells in the body, including cancer cells surgery to remove the polyps and/or cancerous part of the colon.

Interesting Facts about Colon Cancer

People who have FAP can develop hundreds and even thousands of polyps in their colon, whereas people with HNPCC develop relatively few.

The progression from a benign to a malignant cancer typically requires multiple mutations that allow cells to acquire new and abnormal characteristics, such as an increased growth rate, inability to adhere or stick to neighbouring cells, propensity to migrate to other places in the body, etc. For example, at least seven mutations are required to produce a malignant colon tumour.

Inherited cancers often provide clues about the genes mutated in noninherited (sporadic) cancers. For example, mutations in the APC gene are found not only in FAP tumors but in 85% of all sporadic colon tumors as well.

Breast Cancer and Ovarian Cancer

What is Breast Cancer and Ovarian Cancer?

Cells normally grow and divide only when they are needed to keep our bodies functioning properly. But sometimes, the mechanisms that regulate cell growth stop working and cells divide out of control to form tumors. This is called cancer. When cancer develops in the cells of breast or ovarian tissue it is called breast or ovarian cancer, respectively.

Most people who develop breast or ovarian cancer have no history of the disease in their family. In fact, only 5 to 10 percent of all breast and ovarian cancers are caused by inherited genetic factors. These rare cases typically result from inherited mutations in either the BRCA1 or BRCA2 gene.

The BRCA1 and BRCA2 genes are called tumour suppressor genes because they prevent uncontrolled cell growth. BRCA1 is located on chromosome 17, and BRCA2 is located on chromosome 13. Scientists believe BRCA1 and BRCA2 work by fixing damaged or broken DNA. Women who inherit a mutated copy of the BRCA1 or BRCA2 gene accumulate broken and deformed chromosomes, and therefore have a greater chance of accumulating mutations that will lead to uncontrolled cell growth and cancer. Men who inherit the defective genes are also more likely to develop breast and/or prostrate cancer. (Yes, men can get breast cancer.)

How do you get Breast or Ovarian Cancers?

High-risk families include those who have inherited a mutation in either the BRCA1 or BRCA2 gene. The mutated BRCA1 and BRCA2 genes are inherited in an autosomal dominant pattern. A child only needs to inherit one copy of the mutated gene to have an increased cancer risk. Children who have a parent with the BRCA1 or BRCA2 mutation have a 50 percent chance of inheriting the mutation.

Just because a person inherits the defective gene does not mean he or she will develop cancer, but inheritance greatly increases the risk. Out of every 100 women who inherit a mutated BRCA1 or BRCA2 gene, as many as 60 will develop breast cancer by age 50; by age 70, approximately 80 will develop breast cancer.

How do Doctors Test for BRCA1 and BRCA2 Mutations?

A person with a strong family history of breast or ovarian cancer is a likely candidate for genetic screening. By analyzing a sample of the patient's blood, doctors can identify whether the person has inherited the BRCA1 or BRCA2 mutation. The test cannot tell if or when the person will develop breast or ovarian cancer; it can only tell if he or she is at risk because of the faulty gene.

What can a Woman do if she Inherits these Mutations?

Some women who discover that they've inherited the mutated BRCA1 and BRCA2 genes undergo special treatments to protect themselves from breast and ovarian cancer. They may have one or both breasts removed (called a mastectomy or double mastectomy), and their ovaries removed.

They may take the medicine tamoxifen, which is believed to protect against breast cancer.

Interesting Facts about Breast and Ovarian Cancer

The progression from a benign to a malignant cancer typically requires multiple mutations that allow cells to acquire new and abnormal characteristics, such as an increased growth rate, inability to adhere or stick to neighbouring cells, propensity to migrate to other places in the body, etc. Genes involved in the repair of DNA damage (such as BRCA1 and BRCA2) are often associated with cancer. This is because they allow mutations to accumulate at a much faster rate.

Usually, people inherit one of several hundred different mutations of the BRCA1 and BRCA2 genes. But for some reason, Eastern European (Ashkenazi) Jews seem to inherit only three of theses different mutations. Ashkenazi Jews are also 10 times more likely to have mutations in the BRCA1 and BRCA2 genes than any other ethnic group.

Alzheimer's Disease

What is Alzheimer's Disease?

Alzheimer's is a disease that causes dementia, or loss of brain function. It affects the parts of the brain that deal with memory, thought, and language. The brain of a person with Alzheimer's contains abnormal clumps of cellular debris and protein(plaques) and collapsed microtubules (support structures of the cell). Microtubule disintegration is caused by a malfunctioning protein called tau, which normally stabalizes the microtubules. In Alzheimer's patients, tau proteins instead cluster together to form disabling tangles. These plaques and tangles damage the healthy cells around them. The brain also produces smaller amounts of neurotransmitters (acetylcholine, serotonin, and norepinephrine), chemicals that allow nerve cells to talk to one another.

The most common form of the disease, which strikes after age 65, is linked to the apolipoprotein E (apoE) gene on chromosome 19. Scientists don't know how apoE4 increases the risk of developing Alzheimer's. They do know that everyone has apoE, which comes in three forms. One of the forms (apoE4) increases a person's risk of developing Alzheimer's. The other two forms seem to protect against the disease. While people who inherit the apoE4 form of the gene are at increased risk for the disease, they will not necessarily develop it.

Mutations in genes found on chromosomes 1, 14, and 21 are linked to rarer forms of the disease, which strike earlier in life.

How do People get Alzheimer's Disease?

Scientists don't know exactly how people develop Alzheimer's, but they believe it is caused by a combination of genes and environmental factors (multifactorial disorder). The early-onset forms of Alzheimer's are inherited in an autosomal dominant pattern, which means that only one parent has to pass down a defective copy of the gene for their child to develop the disorder.

What are the Symptoms of Alzheimer's Disease?

Because Alzheimer's destroys brain cells, people who have the disorder slowly lose their ability to think clearly. At first, they may forget words or names, or have trouble finding things. As the disorder worsens, they may forget how to do simple tasks (such as walking to a friend's house or brushing their hair). Some people with Alzheimer's also feel nervous or sad.

How do Doctors Diagnose Alzheimer's Disease?

There is no one test for Alzheimer's. Doctors use several different tests to check a patient's memory, language skills, and problem solving abilities. These tests don't diagnose Alzheimer's, but they can rule out other disorders that have similar symptoms.

How is Alzheimer's Disease Treated?

There is no cure for Alzheimer's, but a few medicines can slow its symptoms. A drug called Aricept increases the amount of the neurotransmitter acetylcholine in the brain. A new medicine, Namenda, protects brain cells from a chemical called glutamate, which can damage nerve cells. Doctors may also give their Alzheimer's patients antidepressants or anti-anxiety medicines to ease some of their symptoms. People with Alzheimer's often need a caregiver - someone to help them get around and do the things they were once able to do themselves.

Interesting Facts about Alzheimer's Disease

Alzheimer's was named after the German doctor, Alois Alzheimer, who first named the disorder in 1906. The older a person gets, the higher his or her risk of getting Alzheimer's. Only about 1 or 2 people out of 100 have Alzheimer's at age 65; whereas, one out of every five people has the disorder by age 80.

Down Syndrome

What is Down Syndrome?

Down syndrome is a developmental disorder caused by an extra copy of chromosome 21 (which is why the disorder is also called "trisomy 21"). Having an extra copy of this chromosome means that each gene may be producing more protein product than normal. Cells seem to tolerate this better than having not enough protein, or having altered protein due to a mutation in the DNA sequence. However, producing too much protein can also have serious consequences, as seen in Down syndrome. Genes on chromosome 21 that specifically contribute to the various symptoms of Down syndrome are now being identified.

How do People get Down Syndrome?

Down syndrome is typically caused by what is called nondisjunction. If a pair of number 21 chromosomes fails to separate during the formation of an egg (or sperm), this is referred to as nondisjunction. When that egg unites with a normal sperm to form an embryo, that embryo ends up with three copies of chromosome 21 instead of the normal two. The extra chromosome is then copied in every cell of the baby's body. Interestingly, nondisjunction events seem to occur more frequently in older women. This may explain why the risk of having a baby with Down syndrome is greater among mothers age 35 and older. In rare cases Down syndrome is caused by a Robertsonian translocation, which occurs when the long arm of chromosome 21 breaks off and attaches to another chromosome at the centromere. The carrier of such a translocation will not have Down syndrome, but can produce children with Down syndrome. Javascript is required to view this content.

What are the Symptoms of Down Syndrome?

People with Down syndrome have very distinct facial features: a flat face, a small broad nose, abnormally shaped ears, a large tongue, and upward slanting eyes with small folds of skin in the corners.

People with Down syndrome have an increased risk of developing a number of medically significant problems: respiratory infections, gastrointestinal tract obstruction (blocked digestive tract), leukemia, heart defects, hearing loss, hypothyroidism, and various eye abnormalities. They also exhibit moderate to severe mental retardation; children with Down syndrome usually develop more slowly than their peers, and have trouble learning to walk, talk, and take care of themselves. Because of these medical problems most people with Down syndrome have a decreased life expectancy. About half live to be 50 years of age.

How do Doctors Diagnose Down Syndrome?

Two types of tests check for Down syndrome during a woman's pregnancy: screening and diagnostic tests. Screening tests identify a mother who is likely carrying a baby with Down syndrome. The most common screening tests are the Triple Screen and the Alpha-Fetoprotein Plus. These tests measure levels of certain substances in the blood. Alternatively, ultrasounds (which use sound waves to look inside the mother's uterus) allow the doctor to examine the fetus in the womb for the physical signs of Down syndrome. To confirm a positive result identified in a screening test, one of the following diagnostic tests can be performed: chorionic villus sampling (CVS), amniocentesis, and percutaneous umbilical blood sampling (PUBS). Each takes a sample from the placenta, amniotic fluid, or umbilical cord, respectively, to examine the baby's chromosomes and determine if he or she has an extra chromosome 21. If Down syndrome is not diagnosed in the womb, doctors can usually recognize it after the baby is born by the distinctive facial features. The diagnosis is confirmed with a karyotype - an examination of the baby's chromosomes.

How is Down Syndrome Treated?

No cure exists for Down syndrome. But physical therapy and/or speech therapy can help people with the disorder develop more normally. Screening for common medical problems associated with the disorder, followed by corrective surgery, can often improve quality of life. Moreover, enriched environments significantly increase their capacity to learn and lead a meaningful life.

Interesting facts about Down syndrome

Down syndrome is really the only trisomy compatible with life. Only two other trisomies have been observed in babies born alive (trisomies 13 and 18), but babies born with these trisomies have only a 5% chance of surviving longer than one year.

In 90% of Trisomy 21 cases, the additional chromosome comes from the mother's egg rather than the father's sperm. Down syndrome is the most common genetic disorder caused by a chromosomal abnormality. It affects 1 out of every 800 to 1,000 babies. Down syndrome was originally described in 1866 by John Langdon Down. It wasn't until 1959 that a French doctor, named Jerome Lejeune, discovered it was caused by the inheritance of an extra chromosome 21.

Severe Combined Immunodeficiency

What is Severe Combined Immunodeficiency (SCID)?

SCID is a group of very rare-and potentially fatal-inherited disorders related to the immune system. The immune system normally fights off attacks from dangerous bacteria and viruses. People with

SCID have a defect in their immune system that leaves them vulnerable to potentially deadly infections. There are several types of SCID. The most common form is caused by a mutation in the SCIDX1 gene located on the X chromosome. This gene encodes a protein that is used to construct a receptor called IL2RG (interleukin-2 receptor). These receptors reside in the plasma membrane of immune cells. Their job is to allow two types of immune cells - T cells and B cells - to communicate. When the gene is mutated, the receptors cannot form and are absent from immune cells. As a result, the immune cells can't communicate with one another about invaders in the environment. Not enough T and B cells are produced to fight off the infection, and the body is left defenseless. Another form of SCID is caused by a mutation on chromosome 20 and is characterized by a deficiency of the enzyme adenosine deaminase.

How do people get SCID?

The most common form of SCID exhibits an X-linked recessive pattern of inheritance, and is therefore referred to as X-linked SCID. When a gene is located on the X chromosome, males are more often affected than females. Males do not have a second X chromosome to compensate for the defective one. They only need to inherit one bad copy of the gene to have the symptoms of the disorder. Females, on the other hand, have two X chromosomes. If they inherit one defective X chromosome, they still have its healthy pair. They don't develop symptoms of the disorder, but they still carry the faulty gene and can pass it on to their children.

What are the Symptoms of SCID?

Symptoms usually appear in the first few months of life. Because the immune system cannot protect the baby's body, babies with the disorder tend to get one infection after another. Some of these bacterial infections may be life-threatening, including pneumonia (lung infection), meningitis (brain infection), and sepsis (blood infection).

To make matters worse, SCID patients often don't respond to the antibiotics used to treat bacterial infections. They may suffer more frequently from ear infections, sinus infections, a chronic cough, and rashes on the skin.

Early diagnosis of SCID is very important, because without quick treatment, children with the disease aren't likely to live past age 2.

Bibliography

Alfred Steferud: *Diseases of Fruits and Nuts*, Biotech Books, Delhi, 2005.

Arthur W. Gilbert, Mortier F. Barrus and Daniel Dean: *Growing and Breeding of Potatoes*, Asiatic Pub, Delhi, 2006.

Ashworth S.: *Seed to Seed*, Decorah, Seed Savers Publications, 1991.

Ausubel, F.M. : *Current Protocols in Molecular Biology*, New York: John Wiley and Sons, 1989.

Bahar A. Siddiqui and Samiullah Khan: *Plant Breeding Advances and in vitro Culture*, CBS, Delhi, 1997.

Banki, L.: *Bioassay of Pesticides in the Laboratory,* Akademiai Kiado, Budapest, 1978.

Barbeau, G.: *Tropical Fruits in Nicaragua*, Managua, Nicaragua Ministerio de Desarrollo Agropecuario, Agraria, 1990.

Barnes, N.: *Biology*, New York, Worth Publishers, 1989.

Barnum, Susan R.: *Biotechnology: An Introduction*, Belmont, Thomson/Brooks/Cole, 2005.

Barrington, E. J. W.: *Biochemistry of Primitive Deuterostomians*, London, Academic Press, 1974.

Baruah, Akhil: *Advanced Morphology of Angiosperms*, Aavishkar, Delhi, 2008.

Batlle I. and Tous J.: *Carob Tree (Ceratonia siliqua L.)*. Rome, International Plant Genetic Resources Institute, 1997.

Bauer MW: *Biotechnology-the Making of a Global Controversy*, Cambridge, Cambridge University Press, 2002.

Bhatia, A L : *Biochemistry and Endocrinology,* Indus Valley, Delhi, 2002.

Bhatnagar, Vasudev : *Cell Science and Technology,* Campus Books, Delhi, 2009.

Brandwein, P.F. : *Sourcebook for the Biological Sciences,* San Diego: Harcourt Brace Jovanovich, 1986.

Broach, J.R.: *The Molecular Biology of the Yeast*, Cold Spring Harbor, Cold Spring Harbor Laboratory, 1981.

Brodwin, Paul : *Biotechnology and Culture: Bodies, Anxieties, Ethics*, Bloomington: Indiana University Press, 2001.

Cahill, Lisa : *Genetics, Theology, and Ethics: An Interdisciplinary Conversation*, New York: Crossroad, 2005.

Chaudhary, Vikas: *Entomology and Pest Management*, Navyug, Delhi, 2008.

Chauhan, B.S.　: *Principles of Biochemistry and Biophysics*, Laxmi Publications, Delhi, 2008.

Chiranjib Chakraborty: *Advances in Biochemistry and Biotechnology*, Daya, Delhi, 2005.

Chrispeels, Maarten : *Plants, Genes and Crop Biotechnology*, Sudbury MA, Jones and Barlett Publishers, 2003.

Clark, J.M.: *Experimental Biochemistry*, New York, W.H. Freeman and Company, 1977.

Clawson, Calvin: *The Mathematical Traveller*, New York, Plenum Press, 1994.

Collins, Steven : *The Race to Commercialize Biotechnology: Molecules, Market and the State in Japan and the US.*,　New York: Routledge, 2004.

Collymore L.: *Fruit Production in Barbados*, Port of Spain, Trinidad and Tobago, 1996.

Coste R.: *Coffee: the Plant and the Product*, London, MacMillan, 1992.

Cronquist, A.: *The Evolution and Classification of Flowering Plants*, New York Botanical Garden, Bronx, New York., 1988

Currah L. and Proctor F. J.: *Onions in Tropical Regions*, Kent, Natural Resources Institute, 1990.

Dabholkar, A.R.: *General Plant Breeding*, Concept, Delhi, 2006.

Daphne C. Elliott: *Biochemistry and Molecular Biology*, Oxford University Press, Delhi, 2005.

David Sadava: *Plants, Genes and Crop Biotechnology*, Sudbury MA, Jones and Barlett Publishers, 2003.

Dixon, Dougal: *After Man-A Zoology of the Future*, New York, St. Martin's Press, 1981.

Dodds, John H.: *Plant Genetic Engineering*, New York, Cambridge University Press, 1985.

Doijode S. D.: *Seed Germination in Fruits*, New Delhi, Malhotra Publishers, 1993.

Duckworth, W. L. H.: *Morphology and Anthropology : A Handbook for Students*, Cosmo, Delhi, 2006.

Dudley, E.: *The Critical Villager: Beyond Community Participation*, London, Routledge, 1993.

E. Ramann: *The Evolution and Classification of Soils*, Asiatic Pub, Calcutta, 2006.

Elliott, B N : *Biochemistry and Molecular Biology*, Oxford University Press, Delhi, 2005.

Featherly H. I.: *Taxonomic Terminology of the Higher Plants*, USA, Iowa State College Press, 1954.

Ferentinos L.: *Proceeding of the Sustainable Taro Culture for the Pacific Conference*, Honolulu, HITAHR, 1993.

Fransman M, Junne G, Roobeek A: *The Biotechnology Revolution?*, Oxford, Blackwell, 1995.

Friedberg, E.C.: *DNA Repair*, New York, WH Freeman and Company, 1985.

Fumento, Michael: *Bioevolution: How Biotechnology is Changing Our World*, San Francisco, Encounter Books, 2003.

Ganguly, Smriti : *Biochemistry of Biomolecules*, Pearl Books, Delhi, 2007.

Ghulam Hassan : *Soil Microbiology and Biochemistry,* New India Publishing Agency, Delhi, 2010.

Goodsell, David S.: *Bionanotechnology: Lessons From Nature*, Hoboken, Wiley-Liss, 2004.

Graf, Alfred Byrd: *Advances in Plant Physiology*, Rajat Pub, Delhi, 2008.

Hardy B.: *Biology and Agronomy of Forage Arachis*, Cali, International Centre for Tropical Agriculture, 1994.

Jeffers P.: *Evaluation of Four Onion Varieties in Montserrat*, Plymouth, CARDI, 1992.

Jones, R. M.: *Plant Resources of South-East Asia,* Wageningen, Pudoc Scientific Publishers, 1992.

Khanna, V K : *Objective Genetics, Biotechnology, Biochemistry and Forestry,* I.K. International Publishing House, Delhi, 2008.

Kuppuram, G & K. Kumudamani: *History of Science and Technology in India*, Delhi, Sundeep Prakashan, 1990.

Kurzweil, Ray: *The Age of Spiritual Machines*, New York, Penguin Books, 1999.

Larry V. McIntire : *Biotechnology: Science, Engineering, and Ethical Challenges for the Twenty-first Century*, Washington, DC: Joseph Henry Press, 1996.

Macself, A.J.: *Soils and Fertilizers*, Satish Serial Pub, Delhi, 2005.

Madan Lal Bagdi: *Physiology, Biochemistry and Biotechnology*, Manglam Pub, Delhi, 2007.

Madulid Domingo A.: *A Pictorial Cyclopedia of Philippine Ornamental Plants*, Philippines, Makati Metro Manila, 1995.

Mahindru, S. N.: *Food Safety and Pesticides*, APH, Delhi, 2009.

Meena Francis: *Biotech's Dictionary of Biochemistry*, Biotech Books, Delhi, 2007.

Muneesh Kainth: *Chordate Embryology*, Dominant, Delhi, 2003.

Nobel, P. S.: *Physicochemical and Environmental Plant Physiology*, Academic Press, San Diego, 1999.

Old, R.W. : *Principles of Gene Manipulation*, London, Blackwell Scientific Publications, 1989.

Oldham P.: *Cost of Production of Major Tree Crops in Dominica*, Roseau, Ministry of Agriculture, 1991.

Parry M.L.: *Climatic Change, Agriculture and Settlements*, Dawson Folkestone UK, 1978.

Paul M. Althouse: *Introduction to Agricultural Biochemistry*, Biotech Books, Delhi, 2005.

Pemberton, R. W.: *Predictable Risk to Native Plants in Weed Biological Control*, Oecologia, 2000.

Qystein V. Sjaastad: *Physiology of Domestic Animals*, International Book Distributing Co., Delhi, 2005.

Ragone D.: *Breadfruit: Artocarpus Altilis (Parkinson) Fosberg*, Rome, International Plant Genetic Resources Institute, 1997.

Rifkin, Jeremy: *The Biotech Century*, New York, Penguin Putnam, 1998

Rutherford Lyn.: *A Gourmet's Book of Mushrooms & Truffles*, Sydney, Golden Press Pvt. Ltd., 1991.

Sharma, Pradeep : *Biochemistry and Organisation of Cells*, RBSA Pub, Delhi, 2006.

Stover, R. H. and Simmonds N. W.: *Bananas*, United Kingdom: Longman Scientific and Technical, 1991.

Swarnim, K. : *A Textbook of Biochemistry and Microbiology*, Surendra Pub, Delhi, 2010.

Tawde, A. B.: *Propagation and Rootstocks of Mango*, New Delhi, Malhotra, 1993.

Urton, Gary: *The Social Life of Numbers*, Austin, University of Texas Press, 1997.

Vanangamudi, K.: *Principles and Methods of Plant Breeding*, International Book, Delhi, 2005.

Whealy K.: *The Garden Seed Inventory*, Decorah, Seed Saver Publications, 1988.

White, G.F.: *Natural Hazards: Local, National, Global*, Oxford University Press, New York, 1974.

Woolfe Jennifer A.: *The Potato in the Human Diet*, Cambridge, Cambridge University Press, 1989.

Yadav, M.: *Nutritional Biochemistry and Metabolism*, Arise Pub, Delhi, 2008.

Index

P

Pachytene, 247.
Pantothenic Acid, 50.
Paramutation, 94, 95.
PCR Analysis, 208.
Peptide Bond, 29, 30, 142, 144.
phenol-chloroform, 206.
Photosynthesis, 35, 45, 47, 98.
Plant Mitochondria, 167.
Plastids, 145, 167, 168.
Polyadenylate Polymerase, 164, 167, 168, 169.
Polyadenylation, 73, 138, 162, 163, 164, 165, 166, 167, 168, 169, 171, 175.
Polymerase Chain Reaction, 3, 22, 23, 81, 104, 189, 206, 208.
Prokaryotic Cells, 144, 147, 226, 230.
Prophase, 237, 238, 239, 240, 241, 245, 246, 247, 249.
Protein Degradation, 80.
Protein Folding, 19, 20, 25, 26, 76, 227.
Protein Quantification, 81.
Protein Structure, 19, 20, 37, 67, 69, 229, 239.
Protein Transport, 76.
Proteomics, 33, 35, 37, 38, 39, 40, 87.
Pyridoxine, 51.

Q

Quadruplex Structures, 197.

R

RFLP Analysis, 210, 215.
Riboflavin, 50.
Ribonucleases, 145, 174, 177.

Ribosomal RNA, 72, 135, 141, 143, 144, 145, 150, 157, 160, 222.
Ribozyme, 17, 83, 158, 159, 160, 161, 162.
RNA Tertiary Structure, 14.

S

Saccharides, 56.
Sequencing, 33, 34, 86, 107, 139, 189, 191, 207.
Small Interfering RNA, 178.
Stereochemistry, 223.
STR Analysis, 209, 212, 213.

T

Telophase, 237, 241, 246, 249, 250.
Thiamine, 49.
Topoisomerase, 118.
Transcriptional Bursting, 84.
Transfer RNA, 17, 61, 70, 72, 74, 75, 135, 141, 144, 150, 157, 201, 226.

U

Ubiquitination, 30, 37, 82, 227, 229.

V

Vitamin, 46, 48, 49, 50, 51.

X

X-linked Dominant, 260.
X-linked Recessive, 261.

Z

Zygotene, 247.

❑❑❑